AF771293

FIELDS OF KNOWLEDGE: SCIENCE, POLITICS AND PUBLICS IN THE NEOLIBERAL AGE

POLITICAL POWER AND SOCIAL THEORY

Series Editor: Julian Go

Recent Volumes:

POLITICAL POWER AND SOCIAL THEORY VOLUME 27

FIELDS OF KNOWLEDGE: SCIENCE, POLITICS AND PUBLICS IN THE NEOLIBERAL AGE

EDITED BY

SCOTT FRICKEL
Brown University, Providence, RI, USA

DAVID J. HESS
Vanderbilt University, Nashville, TN, USA

Emerald

United Kingdom – North America – Japan
India – Malaysia – China

Emerald Group Publishing Limited
Howard House, Wagon Lane, Bingley BD16 1WA, UK

First edition 2014

Copyright © 2014 Emerald Group Publishing Limited

Reprints and permission service
Contact: permissions@emeraldinsight.com

No part of this book may be reproduced, stored in a retrieval system, transmitted in any form or by any means electronic, mechanical, photocopying, recording or otherwise without either the prior written permission of the publisher or a licence permitting restricted copying issued in the UK by The Copyright Licensing Agency and in the USA by The Copyright Clearance Center. Any opinions expressed in the chapters are those of the authors. Whilst Emerald makes every effort to ensure the quality and accuracy of its content, Emerald makes no representation implied or otherwise, as to the chapters' suitability and application and disclaims any warranties, express or implied, to their use.

British Library Cataloguing in Publication Data
A catalogue record for this book is available from the British Library

ISBN: 978-1-78350-668-2
ISSN: 0198-8719 (Series)

ISOQAR
REGISTERED
Certificate Number 1985
ISO 14001

ISOQAR certified Management System, awarded to Emerald for adherence to Environmental standard ISO 14001:2004.

INVESTOR IN PEOPLE

CONTENTS

LIST OF CONTRIBUTORS

Mathieu Albert	Wilson Centre, and Department of Psychiatry, University of Toronto, Toronto, ON, Canada
Elizabeth Popp Berman	Sociology Department, University at Albany, SUNY, Albany, NY, USA
Scott Frickel	Sociology Department, Brown University, Providence, RI, USA
David J. Hess	Sociology Department, Vanderbilt University, Nashville, TN, USA
Matthew C. Hoffmann	Sociology Department, Loyola University Chicago, Chicago, IL, USA
Kirk Jalbert	Science and Technology Studies Department, Rensselaer Polytechnic Institute, Troy, NY, USA
Abby Kinchy	Science and Technology Studies Department, Rensselaer Polytechnic Institute, Troy, NY, USA
Daniel Lee Kleinman	Department of Community & Environmental Sociology, University of Wisconsin, Madison, WI, USA
Rebecca Lave	Department of Geography, Indiana University, Bloomington, IN, USA
Jessica Lyons	Public Policy Department, Rochester Institute of Technology, Rochester, NY, USA
Wendy McGuire	School of Social Work, York University, Toronto, ON, Canada

Kelly Moore	Sociology Department, Loyola University Chicago, Chicago, IL, USA
Libby Schweber	School of Construction Management and Engineering, University of Reading, Reading, UK
Sainath Suryanarayanan	Department of Community & Environmental Sociology, University of Wisconsin, Madison, WI, USA

SENIOR EDITORIAL BOARD

Ronald Aminzade
University of Minnesota

Eduardo Bonilla-Silva
Duke University

Michael Burawoy
University of California-Berkeley

Nitsan Chorev
Brown University

John Coatsworth
Columbia University

Diane E. Davis
Harvard University

Susan Eckstein
Boston University

Peter Evans
University of California-Berkeley

Nora Hamilton
University of Southern California

Eiko Ikegami
New School University Graduate Faculty

Howard Kimeldorf
University of Michigan-Ann Arbor

Florencia Mallon
University of Wisconsin-Madison

Jill Quadagno
Florida State University

Ian Roxborough
Stony Brook University

Michael Schwartz
Stony Brook University

George Steinmetz
University of Michigan

John D. Stephens
University of North Carolina-Chapel Hill

Maurice Zeitlin
University of California-Los Angeles

Sharon Zukin
City University of New York

STUDENT EDITORIAL BOARD

Cara Bowman

Zophia Edwards

David Levy

Megan O'Leary

EDITORIAL STATEMENT

Political Power and Social Theory is a peer-reviewed annual journal committed to advancing the interdisciplinary understanding of the linkages between political power, social relations, and historical development. The journal welcomes both empirical and theoretical work and is willing to consider papers of substantial length. Publication decisions are made by the editor in consultation with members of the editorial board and anonymous reviewers. For information on submissions, please see the journal web site at www.bu.edu/sociology/ppst.

SERIES EDITOR'S INTRODUCTION

I am delighted to present this special volume of PPST, edited by Scott Frickel and David J. Hess. The essays in *Fields of Knowledge: Science, Politics and Publics in the Neoliberal Age* track transformations in scientific and technological policies and practices that unfolded with the rise of neoliberal thought since the 1970s. And they do so through deft deployments of field theory and organizational theory, merging the sociology of knowledge with broader theories of social organization and politics. In this way, these essays carry on the PPST tradition of offering empirically grounded and theoretically engaged critical social science. I am grateful to Scott Frickel and David J. Hess for choosing PPST and to the contributors for their insightful essays. I am also thankful to Thomas Dark and the rest of the team at Emerald for putting it all together.

Julian Go
April 2014
Boston, MA

INTRODUCTION: FIELDS OF KNOWLEDGE AND THEORY TRADITIONS IN THE SOCIOLOGY OF SCIENCE

David J. Hess and Scott Frickel

ABSTRACT

This Introduction gives a historical and theoretical overview of this volume on Fields of Knowledge: Science, Politics and Publics in the Neoliberal Age, *which showcases original research in political sociology of science targeting the changes in scientific and technological policy and practice associated with the rise of neoliberal thought and policies since the 1970s. We argue that an existing family of field theoretic frameworks and empirical field analyses provides a particularly useful set of ideas and approaches for the meso-level understanding of these historical changes in ways that complement as well as challenge other theory traditions in sociology of science, broadly defined. The collected papers exhibit a dual focus on sciences'* interfield relations, *connecting science and science policy to political, economic, educational, and other fields and on the* institutional logics *of scientific fields that pattern expert discourses, practices, and knowledge and shape relations of the scientific field to*

Fields of Knowledge: Science, Politics and Publics in the Neoliberal Age
Political Power and Social Theory, Volume 27, 1–30
Copyright © 2014 by Emerald Group Publishing Limited
All rights of reproduction in any form reserved
ISSN: 0198-8719/doi:10.1108/S0198-871920140000027001

the rest of the world. By reconceptualizing the central problem for political sociology of science as a problem of field- and inter-field dynamics, and by critically engaging other theory traditions whose assumptions are in some ways undermined by the contemporary history of neoliberalism, we believe these papers collectively chart an important theoretical agenda for future research in the sociology of science.

The essays in this volume were first presented at a workshop that was held at the University of Wisconsin in June, 2012, and organized by Daniel Lee Kleinman. The workshop brought together a network of scholars who study the political sociology of science and technology, an approach to the sociology of science that draws systematic attention to power dynamics and structures of inequality within research fields and in the "extrafield" relations among science, the state, industry, social movements, and in civil society more generally. Recent research in this tradition has been influenced by the edited volume *The New Political Sociology of Science* (Frickel & Moore, 2006) and by a conference on the political sociology of science held at Rensselaer Polytechnic Institute in 2011 (Hess, 2011). The aim of the 2006 book was to sketch a broad framework for the sociological analysis of power and inequality as they are influenced by and impact science, technology, and society. Toward that end, the volume was general in its topical focus, theoretical scope, and methodological preferences, with authors contributing chapters on a range of issues, concepts, and approaches. The 2011 conference at Rensselaer, designed primarily to take stock of new research in the area, was similarly general in its attention to theory, scope, and method.

By contrast, the papers collected for the present volume of *Political Power and Social Theory* focus on a historically specific problem area: the changes in scientific and technological policy and practice associated with the rise of neoliberal thought and policies since the 1970s (e.g., Moore, Kleinman, Hess, & Frickel, 2011). Four decades on, ascendant neoliberal ideologies and the policies associated with those ideas continue to have major impacts on the practice, organization, and uses of scientific research. In turn, the neoliberalization of science is complicating interactions among states, industries, and civil society.

All of the papers presented in this volume take the historically changing configuration of field-level relationships emergent under neoliberalism as their collective starting point. In doing so, these papers extend current work on neoliberalism and science in important new directions. To date, much of this work has focused on the neoliberalization of the university, with special emphasis on university–industry relationships (e.g., Slaughter & Rhoades, 2004). The research presented in the present volume significantly broadens this perspective, from the impacts of neoliberalism on universities (and university-based science) to newly emergent institutional configurations involving multiple fields that nurture and constrain scientific production in new and often troubling ways.

Such changes often call into question assumptions of scientific autonomy and the entrepreneurial independence of scientists that underlie many of the most influential works in the sociology of science and scientific knowledge. The present volume engages with and builds on existing theory traditions in the sociology of science, such as the focus on the reward system and the normative regulation of the scientific field in the functionalist tradition and on the social negotiation of knowledge claims among networks of scientists in the constructivist tradition. However, in maintaining a wider focus on the dynamic interaction of enduring structures of power and inequality in society with the agency of both scientists and nonscientists as they circulate within and across diverse social fields, authors of the current volume are drawn to and in some cases synthesize a range of meso-level frameworks from field theory, neo-institutional theory, and governmentality theory. By reconceptualizing the central problem for political sociology of science as a problem of field- and inter-field dynamics, and by critically engaging other theory traditions whose assumptions are in some ways undermined by the contemporary history of neoliberalism, we believe these papers collectively chart an important theoretical agenda for future research in the sociology of science.

OVERVIEW OF THE ARTICLES

All of the articles in the volume take as a point of reference some version of field theory, including Bourdieusian, strategic action, and organizational field theories. Some of the researchers find field theory to be helpful for the problems that they address, whereas others have used it more as a foil for the articulation of alternative theoretical frameworks, especially frameworks that draw on neo-institutional theory. The papers fall roughly into two main groups that have to do with ways in which one might conceptualize field theory: the problem of interfield relations and the problem of institutional logics.

Interfield Relations

In the first group of articles (by Albert and McGuire, Lave, Hess, and Suryanarayanan and Kleinman), the researchers focus on a problem in field theory that has been identified as an area that Bourdieu tended to underplay in his writings on science: the study of the relations among fields (Camic, 2011). Although Bourdieu developed a theory of the relationship between the field of power and the intellectual field, there is general sentiment among scholars sympathetic to field theory that it needs to do more work to address issues of interfield relations, and this general issue in field theory becomes acute when one is concerned with the neoliberalization of the scientific field. In "Understanding Change in Academic Knowledge Production in a Neoliberal Era," Mathieu Albert and Wendy McGuire work from within a Bourdieusian field theory framework to develop the concept of the autonomous–heteronomous polarity of the scientific field. They use the changing weights of the producer-oriented and user-oriented poles of the scientific field to modify the "Mode 2" thesis that scientific research has increasingly shifted toward mission-based research (Gibbons et al., 1994). In contrast, Albert and McGuire argue that "Mode 1" and "Mode 2" science operate simultaneously and that individual scientists negotiate their relationship with both (Gibbons et al., 1994). Tracking changes in the research funding priorities of the Canadian government in favor of agendas consistent with the country's economic strategy, they draw on Bourdieu's concept of the convertibility of capital to suggest the fluidity with which scientists may move between a user-oriented pole of research to a producer-oriented pole. Although many scientists in their sample adopted a strategy consistent with the traditional vision of science as production for producers, Albert and McGuire found significant generational differences in how scientists responded to the changes, with younger scientists more prone to employ a hybrid strategy. There were also differences between the vision or ideal of the scientists and their actual practices, and in some cases scientists grappled with a misalignment between the vision and practice.

Like Albert and McGuire, Rebecca Lave also uses Bourdieusian field theory for her study of regulatory policy and science for stream restoration, and she builds on her book (Lave, 2012), which also used the distinction between the heteronomous and autonomous poles of the scientific field. In the essay in this volume titled "Neoliberal Confluences: The Turbulent Evolution of Stream Mitigation Banking in the US," Lave develops an analysis of both interfield relations (the stream mitigation industry, regulators, Congress, scientists, and the administration) and intrafield relations within the regulatory field between mitigation bankers and regulators. Stream mitigation banking is a market-based practice that allows people who relocate or culvert a stream to make amends by restoring a stream elsewhere. The growth of stream mitigation banking as an industry has shaped the agendas of the research field, which developed a new science of ecosystem services. Thus, Lave argues that the market-oriented form of regulation created a new industry and science, but she also notes that the regulatory policy is a hybrid of market-oriented, neoliberal elements with older command-and-control approaches to environmental policy. Although not originally developed with a neoliberal intent, the valuation of ecosystem services facilitated the commodification of ecosystems that was needed to develop stream mitigation markets.

In "Beekeepers' Collective Resistance and the Politics of Pesticide Regulation in France and the United States," Sainath Suryanarayanan and Daniel Lee Kleinman study the different regulatory outcomes to the problem of accelerated honey bee die-offs in France and the United States. At one level, the study is a comparison between the more social liberal policy regime of the French government, in which the state is more insulated from industrial influence and more interventionist in economic matters, and the more neoliberal regime of the United States, in which industrial influence on regulatory policy is generally stronger and government regulation of industry is often weaker. However, they reject simplistic formulations that would explain the difference as a result of general differences in political culture, and instead they approach the comparative question through the analysis of interfield relations that takes into account the relations among industrial firms, scientists, regulators, beekeepers, and farmers. They argue that the more interventionist regulatory response to the problem of pervasive honey bee deaths in France is due to a confluence of higher social movement mobilization, different epistemic cultures of honey bee biology, and different structures of government that enable the state to be more or less responsive to movement mobilization. The French beekeepers were more highly mobilized and unified than their American counterparts, but Suryanarayanan and Kleinman suggest that the effectiveness of the mobilization depended in part on differences in epistemic cultures for gauging honey bee health. In France, there was greater willingness to accept the results of laboratory studies as a sound basis for a precautionary policy, whereas in the United States industry prevailed with the view that only controlled field studies produced definitive knowledge and that laboratory studies were inadequate because they did not reflect real-world conditions. In turn, the structure of government interacted with the other two explanatory factors: the more centralized and hierarchical structure in France made it easier for authorities to respond quickly to popular mobilization, and the more fragmented structure in the US government favored policy stasis.

In "When Green Became Blue: Epistemic Rift and the Corralling of Climate Science," David J. Hess also addresses the problem of interfield relations by examining a case in which the standard circulation or "metabolism" of science and the state breaks down. In parallel with discussions of ecological rift in environmental sociology, Hess argues that the rise of climate denialism among elected political officials in the United States has led to an "epistemic rift" in which a relatively strong consensus among a research community is no longer accepted as the politically neutral epistemic doxa, that is, as the basis of "what is the case" upon which normative debate about "what should be done" can take place. The breakdown occurs because a threatened industry has funded scientists and media pundits who attack of scientists and media pundits who attack the consensus of the research field that contemporary climate change is fundamentally shaped by anthropogenic greenhouse gases. Thus, the scientific research community is "corralled" or cut off from its public education and public policy functions, political officials can openly pronounce anti-science views without losing legitimacy at the polls, and, once in office, the denialist politicians can also target climate science research for defunding. Under these circumstances, the scientific community becomes part of the counterpublic of a coalition of environmentalists, green businesses, and political progressives who advocate continued reform.

In summary, this group of four studies suggests some new ways to think about science from the perspective of interfield relations. As Albert and McGuire note, there is continued value in developing the heteronomous–autonomous pole concept from Bourdieu and for understanding how scientists negotiate changes in the valorization of the polarities of a research field, and as Lave argues, there is also value in viewing regulatory policy from a field perspective and in relationship to emergent sciences. Suryanarayanan and Kleinman also suggest the value of a comparative analysis of the interfield relationship in the sociology of science, with attention to the relations among social movements, epistemic cultures, and characteristics of the state. Hess uses the negative case of the breakdown of traditionally understood interfield relations to develop an explanation of the conditions under which epistemic rift and the corralling of science occur.

Institutional Logics of Fields

The second group of papers works with another problem that has sometimes been identified in field theory: the need to develop a level of cultural analysis that goes beyond the concepts of habitus and skills in Bourdieusian and strategic action versions of field theory. The authors in this group have mostly turned to neo-institutional theory, which in its early formulations borrowed heavily from Bourdieusian field theory, and within neo-institutional theory they focus on the concept of institutional logics. In "The Cultural Role of Science in Policy Implementation: Voluntary Self-Regulation in the UK Building Sector," Libby Schweber focuses on green building standards in the United Kingdom as an example of the liberalization of industrial regulation through voluntary standards. She draws on the analysis of organizational fields in sociological neo-institutionalism, and she is particularly interested in multiple institutional logics, such as the contrast between environmentalism and commercialism, which firms and other actors in the organizational field must negotiate. She then makes the bold theoretical move to synthesize neo-institutional theory with governmentality theory by arguing that institutional logics are embedded through multiple mechanisms elucidated by governmentality theory (e.g., visibilities, knowledge, techniques and practices, and identities). In addition to extending neo-institutional theory by studying the micro-dynamics of techniques and instruments (rather than rules and sanctions), she also argues for an expansion of the focus of the new political sociology of science from the role of science in policymaking to its role in policy implementation.

In "Field Theories and the Move toward the Market in US Academic Science," Elizabeth Popp Berman examines the theory of Fligstein and McAdam (2011, 2012), who draw on Bourdieu, organizational field theory, and social movements theory to develop a theory of strategic action fields. One of the advantages of their approach, she notes, is its explicit emphasis on interfield relations, but she also draws out some shortcomings based on her analysis of the changes in academic science in the United States since the 1970s. She notes that in any field there will always be multiple institutional logics available, but some prevail over others (i.e., the market logic over the science logic in academic science during the period of neoliberalization). One factor that affects the change in fortunes of an institutional logic is the environmental conditions of the field, a factor that is emphasized in Fligstein's and McAdam's discussion of exogenous shocks as a factor that leads to episodes of contention between challengers and incumbents (similar to the emphasis on institutional entrepreneurs in neo-institutional theory). However, Berman suggests that challengers within the scientific field were not the primary driver of field change; instead, changes were shaped by the shift of policymakers' preferences toward research that met broader political goals of innovation and economic competitiveness. In turn, the changes in policy direction were shaped in part by representatives of large companies that had substantial research-and-development operations and were pushing policymakers to alter the policy landscape in favor of innovation research. She argues that although some episodes of field change do involve dramatic instances of contention led by challengers within the field who may take advantage of exogenous shocks, the case of the transformation of academic science in the United States was a more gradual process that is characterized by ongoing contention and instability. Thus, she points to the need to consider various pathways toward field change, of which the Fligstein and McAdam model of exogenous shocks followed by a mobilization by a challenger is only one possibility.

In "'The Tip of the Day': Field Theory and Alternative Nutrition in the US," Kelly Moore and Matthew C. Hoffmann study the circulation of knowledge associated with nutritional science to suggest a method for studying field change that attends to multi-decade changes in the relations among government policy, the food and health industries, and social movements. Finding Bourdieu's focus on capital too confining, they turn instead to neo-institutional theory and its openness to a broader set of motivations and multiple institutional logics. They then review the history of nutritional policy in the United States to show how the rise of the Civil Rights movement led to new concern with hunger among the poor, especially African Americans. The concern led to antihunger policies as well as shifts in the research agendas of the nutritional science field toward issues related to hunger and malnutrition. However, by 1977, the field of nutrition policy had changed again to focus on the problem of overconsumption, and new models began to circulate that emphasized individual responsibility in the face of chronic disease and overconsumption. At the same time, social movements and an emerging wellness industry emphasized the advantages of individual decisions that would lead to a healthier lifestyle, and the field of nutritional research began to shift toward issues involving chronic disease and health. Through detailed case studies of two examples of nutritional experts who also wrote for a popular audience, Moore and Hoffmann show how these hybrid scientist-popularizers blend institutional logics from biomedicine, American self-reform traditions, and neoliberal entrepreneurialism. Like Berman, who also focused on the problem of multi-decade change, Moore and Hoffmann draw attention to shifting institutional and cultural logics and changing interfield relations, and they also highlight the importance of the circulation of scientific knowledge from academic researchers through policy regimes, industrial change, and social movements.

In "What is Volunteer Water Monitoring Good for? Fracking and the Plural Logics of Participatory Science," Abby Kinchy, Kirk Jalbert, and Jessica Lyons examine the institutional context in which citizen science has risen to prominence as a method of investigating the impacts of shale gas development by hydraulic fracturing or "fracking." Now widespread in the United States, volunteer watershed monitoring reflects, on one hand, broader neoliberal shifts to cut government costs by increasingly relying on civil society organizations and volunteers to provide important services, and, on the other hand, public demands for increased participation in environmental science and decision-making. Three institutional logics have historically shaped volunteer water monitoring practices: popular education, science, and environmental policing. However, the introduction of fracking – an exogenous shock – has driven multiple changes in the interdependent fields of academic science, environmental regulation, and participatory water monitoring, leading to the mixing and reordering of these logics. In comparison with many water monitoring projects of the past, water monitoring organizations are adopting practices from academic science and environmental advocacy organizations that reflect the scientization of the regulatory field as well as the pressures on academic scientists to remain politically neutral.

In summary, the second group of four articles develops a neo-institutional approach to the scientific field by examining the changing relations among institutional logics that elucidate long-term changes in a scientific field. Schweber suggests the value of governmentality studies to enable a concrete connection between broad institutional logics and specific techniques and practices, and Berman argues in favor of a multiple pathways approach to changes in institutional logics. Thus, whereas the Fligstein and McAdam approach of exogenous shocks and relatively short-term episodes of contention led by challengers may characterize some examples of field change, she argues that the study of field change should explore other pathways, including ones based on policy-driven change that, at least in her case, was partially instigated by industry. Moore and Hoffmann also take a long-term perspective on changes in the scientific field, and they draw attention more to the circulation of scientific knowledge through the state, industry, and social movements and the multiple cultural logics at play. Kinchy and colleagues also attend to multiple institutional logics, and they show how a relatively depoliticized field of citizen science has changed in response to the threat of hydraulic fracturing, the lack of government response to citizen concerns, and the mobilization of activist organizations.

As a group, the eight essays suggest the integrated nature of a next generation of field and neo-institutional approaches to the sociology of science. The focus on interfield relations inevitably involves multiple institutional logics because of actors who move across fields or interact with actors in other fields. These institutional logics are not bounded by agents or by fields, but they tend to transect them. However, in any specific point in the history of a field, some institutional logics are more and less dominant. Thus, they do not float like a superorganic entity above social relations and social structure, and they require sociological analysis and explanation.

INSTITUTIONS, NORMS, AND INTERESTS

Collectively, these articles make the general argument that there is a role for theoretical frameworks in the sociology of science that address the issue of political power and draw on theory traditions associated with field analysis, neo-institutional analysis, governmentality studies, and social movement studies. The articles also make a more specific argument about the need to pay greater attention to interfield relations and institutional logics within these theory traditions. But how do these arguments fit into the broader social theory traditions in the sociology of science and the interdisciplinary conversations of STS (science and technology studies)? The history of theory traditions within the sociology of science (a term used here in the broad sense, to include the sociology of scientific knowledge) suggests that the fit may be uncomfortable but also potentially generative.

Because several of the essays make reference to institutional theory, a good starting point for an answer to the question is to review the historical basis of the concept of the institution and institutional analysis in the sociology of science. Although Max Weber makes frequent reference to institutions, Talcott Parsons is generally credited with developing the social theory of institutions, and his theory was in turn indebted to various predecessors, including institutional economists such as Thorstein Veblan. Charles Camic uncovered an unpublished early manuscript by Parsons on the topic, and the manuscript and several commentaries, including one by Camic (1990), appeared in a 1990 issue of the *American Sociological Review* (Parsons, 1990). In the essay, Parsons rejected behavioralist approaches and linked the general theory of institutions to normative rules that were associated with sanctions and moral authority. In the subsequent 1942 essay on "Propaganda and Social Control" (Parsons, 1954), Parsons developed the analysis of institutions by situating it in the context of the question of Japanese and German military aggression. He described institutions as "patterns governing behavior and social relationships" that are based on a system of statuses and roles, the associated norms that govern behavior, and the broader cultural traditions of which the norms are a part. Adherence to norms in turn depends on a combination of a system of positive and negative sanctions and on the state of mind of the actor, which includes a belief in the legitimacy of the norms.

In the same year, Robert Merton published a parallel essay originally title "Science and Technology in a Democratic Order" that was later retitled "The Normative Structure of Science" (Merton, 1973). Also situated in the context of World War II, the essay was concerned with the need for the autonomy of science as an institution and the importance of that autonomy for the maintenance of a democratic society, with Soviet and Nazi science looming as counter-examples. Although the primary concern at the time was fascism, if one substitutes his concern with the domination of the scientific field by authoritarian politics with the current problem of the domination of the scientific and political fields by large corporations, in many ways the essay is quite contemporary. Like Parsons, Merton's institutional analysis drew attention to norms and values, but for the case of science as an institution he recognized the importance of both technical norms (e.g., logical consistency, valid prediction) and moral norms. The latter included the much-debated list of universalism, communism, distinterestedness, and organized skepticism. By the time of his 1957 paper "Priorities in Scientific Discovery: A Chapter in the Sociology of Science," which was based on his presidential address before the American Sociological Association, Merton had made several important changes in his thinking about science as an institution. First, his discussion of priority disputes led to an emphasis on the norm of originality, which was more closely tied than the other norms to the reward system, and he focused on how the reward system bestows recognition from one's peers on work deemed original and important. This point was developed later in

Hagstrom's discussion (1965) of science as based on competition for recognition, rather than on the lingering effects of the socialization of norms, and in Bourdieu's analysis (1975) of the particular form of capital that, as in other intellectual fields, can only emerge from the field itself. But Merton also recognized that the norm of originality could lead to dysfunctional struggles over priorities, and hence he suggested the existence of a counter-norm of humility, with the resulting condition of sociological ambivalence. The most well-known of the conflicts between norms and counter-norms, the issue of the extent to which universalism or particularism better describes the behavior of scientists and the functioning of science as a social institution, led to a fruitful, long-term competition of research programs, between Merton's network and a group of other American sociologists, that is sometimes known as the "fair science" controversy (see Hess, 1997). The controversy also led to increasing recognition of the difficulties of applying the analysis of norms and counter-norms to describe episodes in the history of science (e.g., Mitroff, 1974). For Merton the shift to the analysis of the reward system, rather than the conflicts among norms in the decisions of individual scientists, led to a fruitful research program that included his influential work on cumulative advantage theory. In turn, the discussions of cumulative advantage, particularism and universalism, and other aspects of the fairness of science, shared affinities with the broader political programs of social liberalism, such as concern with equality of opportunity and with corrective policies, such as affirmative action, that improve universalism (Hess, 2013).

Although Merton had already recognized the difficulties of applying norms to the actions and decisions of individual scientists, an influential and more systematic critique of norms came from the emerging networks of the sociology of scientific knowledge. Barnes and Dolby (1970) argued that the analysis of norms was troubled by the inconsistencies among a wide range of scientific practices across time periods, countries, and disciplines, and they suggested that a better approach would rely on a methodology based on Kuhn's paradigm concept (1970) but carefully configured to allow for the clash of different paradigms among different "consensus groups." Thus, they recognized the crucial role of competing networks that in some senses predated field theories of science. They also argued that under the circumstance of scientific controversy, professions of norms (both technical and social) may be commonplace, but Barnes and Dolby saw those statements as part of the polemical context. In effect, their shift to the importance of paradigms meant the excision of social norms from the analysis of science and, in Mertonian terms, the focus more exclusively on technical norms that were embedded in specific disciplines, subdisciplines, and paradigms.

Mulkay (1976, see also Mulkay & Gilbert, 1982) developed a similar argument based on the empirical claim (especially developed in Mitroff, 1974) that scientists in fact frequently deviate from general social norms such as rationality, emotional neutrality, universalism, distinterestedness, impartiality, communality, humility, and organized skepticism. Instead, the only relevant norms are "furnished, not by this set of social norms, but by the scientific frameworks and technical procedures in terms of which the research community is internally differentiated" (1976, p. 639). Mulkay then drew on historical work on the political context of the articulation of the pure science ideal (Daniels, 1967) to develop the argument that social norms were in fact an occupational ideology of scientists used to justify their autonomy and to support their relations with the state or, to use our terminology, the extrafield patrons of science. Drawing on ethnomethodology, Mulkay argued that norms are repertoires that scientists select from and perform based on their social interests. In effect, they are outcomes of scientists' action rather than causes. This agent-centered approach to the analysis of science was then taken up in many instances in the subsequent history of the sociology of scientific knowledge, including more recent work on performativity and economics (e.g., Callon, 2007).

More generally, a consensus emerged that the analysis of social norms was inadequate for the emerging task of a sociology of scientific knowledge (Bloor, 1991), which studied the "content" of knowledge (such as the controversies over theories, methods, and empirical claims) rather than the institutions of knowledge production (e.g., the reward system). However, although Mulkay had suggested the potential for an analytical strategy grounded in the interests of scientists, similar problems occurred with interests analysis, the first main research program in the sociology of scientific knowledge in the strong-program tradition. The problems occurred mainly with the attempt to link controversies between networks within a research field (such as rival groups of statisticians, where an "interest" in a network's success was visceral) to broader social statuses and political ideology, where connections with broader social interests were more problematic (MacKenzie, 1981, 1983, 1984; MacKenzie & Barnes, 1979). These studies provoked a firestorm of criticism, again suggesting the tenuousness of sociologies of scientific knowledge that attempt to forge a linkage to extrascientific statuses, and the complaints about scientists as "interest dopes" were memorialized in Woolgar's phrase, "instead of norms we now have interests" (1981, p. 375).

It might have been possible for MacKenzie to find archived statements by the statisticians he studied that showed, in their own minds, the connections between their intellectual positions within the scientific field and their broader social and political positions. But the difficulty of finding such needles in haystacks helped to motivate the turn toward contemporary ethnographic methods in studies of laboratories and controversies. Of these studies, Harry Collins's work was one of the most sociological research programs, and his empirical program of relativism (which was also the precursor of the social construction of technology program in technology studies) recognized the value of relating the mechanisms of closure of a controversy to "the wider social and political structure" (1983, p. 96). In his description of this third phase, he suggested that alignments between one side of a controversy within the scientific field with positions outside the field (such as industrial power) could explain how one side of a controversy had access to greater resources that helped it to close a controversy in its favor. However, this methodological insight remained largely undeveloped in subsequent work in the sociology of scientific knowledge.

Whereas Collins's method suggested a way to solve the problem of imputation of interests through a causal analysis of the effects of extrafield alignments with resource availability, there was also a solution to the problem of the third phase through cultural analysis. Traweek's ethnographic study (1988) of physicists was exemplary in this regard, and one really needs to follow her whole corpus (including work on women Japanese physicists) to see the import of the method that was being developed in cultural anthropology and feminist cultural analysis. Cultural analysis could link extrafield statuses (i.e., the status of physicists as Japanese or American, men or women, and so on) to technical distinctions within the field (preferences for research programs and detector designs) without having to run the gauntlet of causal explanation. In other words, through the analysis of "cultural logics," or homologies of difference (Japanese is to American as one type of physics is to another), one could bring to light the alignment of distinctions in the broader social structure (nationality, race, gender, class, political affiliation) and distinctions in technical culture without making claims that scientists intentionally imported an allegiance to a specific extrascientific position into their intellectual positions in the scientific field. Interfield homologies, like other semiotic structures, may be visible only to the social scientist, but as noted above one must exercise caution in the analysis so as not to postulate relationships without some empirical evidence. Thus, it is important to pay attention to shifters, to mediating categories, that are identical and enable translation across different domains. Bourdieu (1981) used a similar method in his studies of misrecognition and the analysis of Heidegger's philosophy, and he added that the method was necessary to avoid the "short-circuit effect" of interests analyses associated with the early applications of the strong program (1990). Thus, cultural analysis (a term that Bourdieu did not use but essentially was what he was doing in his analysis of Heidegger and the philosophical field) could accomplish the goals of interests analysis and Collins's third phase without treading into the murky water of the explicit motivations of individuals (Hess, 1995).

Cultural analysis of one form or another became widespread in the STS field (at least if one defines the boundaries of the field broadly), and it was especially prominent in work by anthropologists and feminists (e.g., Haraway, 1989; Harding, 1998; Martin, 1994; Rapp, 2000), but it never took root in the sociology of scientific knowledge. Traweek's work suffered a dismissive review by Latour (1990), who argued that she was part of a tradition in anthropology running from Durkheim and Mauss through Lévi-Strauss, who all took for granted the distinction between nature and culture that his analysis of hybrids avoided. Although perhaps convincing to those unfamiliar with the anthropological tradition that he referenced, to those more familiar with the research tradition the argument misses the more complicated discussion of the co-constitution of natural and social categories that emerged in Lévi-Strauss's *Totemism* (1963) and was widely recognized in the field, including in many analyses that applied anthropological structuralism to modern society (e.g., Sahlins, 1976). In any case, Latour's review had the effect of dampening the engagement between cultural anthropology and the sociology of scientific knowledge. It was not until the thirty-year retrospectives of laboratory studies at the annual meeting of the Society for the Social Studies of Science that Traweek earned the field's recognition in the form of a place at the table of a plenary section. But in general, the two fields went their separate ways, and from comments that we heard during the 1990s, the influx of anthropologies, feminisms, and cultural studies into the STS field during that decade was perceived as a threat.

Another "problem" (depending on one's viewpoint) with cultural analysis was the ease with which it could be turned back onto the history of the sociology of science in the form of a critical, reflexive sociology. Thus, one might ask why the turn toward individualized, agency-based frameworks, of which actor-network theory was the epitome, took root during the 1970s and 1980s. The beginning of an answer to the question might take advantage of the comparative perspective of changes that were going on in other fields as well, such as the turn toward ritual and performance in cultural anthropology (Geertz, 1973) and toward resource mobilization theory in social movement studies (McCarthy & Zald, 1973). It is not hard to develop an analysis of the cultural homologies between the rise of agency-based theories in the social sciences and the turn toward individual responsibilization described by Foucault (2008) in his studies of neoliberalism (see also Hess, 2013; Mirowski & Nik-Khah, 2007; Moore & Hoffmann, this volume). This is not to say that any particular defender of an agency-based theory in the social sciences had neoliberal political leanings (a claim that would fall victim to the "short-circuit" problem), or that political sociologists are all socialists. In fact, participation in the STS field for decades has taught us that there is a striking similarity of political positions among social scientists in the field, with most endorsing some form of social liberalism. But cultural analysis is interested in another level, the homologies of distinctions between political ideologies and social theory positions that can be related through common grounds (the shifters), such as the focus on individual action.

Despite the opportunity offered by cultural analysis to develop one type of a sociology of scientific knowledge that takes into account extrafield relationships, the focus of most of the early research programs in the sociology of scientific knowledge remained on the microsociological problem of how networks of scientists negotiate contending claims about theories, methods, research results, and agendas for future research. This focus was, like all research programs, both productive and restrictive; the sociology of science has benefited from a wide range of highly useful concepts such as interpretive flexibility, experimenter's regress, closure, enrollment, and obligatory points of passage. However, in the strongest versions of this tendency, the work became an anti-sociology in which social structure was viewed as an epiphenomenon of the action of individuals or heterogeneous networks, and the emergence of such positions occasioned a division within the sociology of scientific knowledge (Bloor, 1999; Yearley, 2005). The pitched nature of these debates tended to subside over time, especially after the field shifted toward studies of sciences and publics and began to engage more with questions of power and inequality in the circulation of knowledge among expert communities and their counterparts in the state, industry, and the media.

This brief review is helpful to situate the potential reception of a term such as "neo-institutional" analysis within the context of the history of the sociologies of science. Given the freighted history of concepts such as norms and interests in the sociology of scientific knowledge, there is a tendency in the field to police their reemergence. For example, we witnessed a young naïf who, at an annual meeting of the Social Studies of Science, invoked an interests explanation, and a senior figure quickly rose to remind the audience that the field views interests as outcomes of action rather than causes of it and that such issues were settled long ago. Likewise, an article in the journal *Social Studies of Science* criticized sociologists who advocate the inclusion of structural explanation in the analysis of scientific change (e.g., Frickel & Moore, 2006; Kleinman, 2003), argued that "coherent cultural and institutional logics are nowhere to be found" (Jones, 2009, p. 843), and concluded that Mertonian norms were still a dead horse. In a subsequent article published in the same journal that might be read as a nonpolemical rebuttal, Suryanarayanan and Kleinman (2013) retained the discussion of norms and showed how research norms (or "technical norms" in terms of the discussion above) such as a preference for false negatives came to dominate the research field that links pesticides to colony collapse disorder. The authors also showed that the dominance of the norms led to the marginalization of the epistemic claims of commercial beekeepers, and they added, "The ways in which honey bee scientists approached insecticide issues were influenced by the broader agricultural organizational and scientific communities within which they were situated, where the dominant agro-entomological knowledge framed insects as pests that they needed to be targeted through chemicals" (2013, p. 14). Here, then, is an example of delivery on the deferred promise of Collins's third phase, but without the problematics of social norms and the short-circuit of Edinburgh interest analysis. Suryanarayanan and Kleinman use the concept of norms to draw out a comparative analysis of differences within a research community and its effects on the status of lay knowledge in the interfield relationship among bee researchers, pesticide manufacturers, regulators, farmers, and beekeepers.

Although some researchers no longer use the term "norms," even technical norms, due to the history outlined above and the risk of being accused of Mertonian atavism, a family of related concepts continues to emerge. Like the return of the repressed, there appears to be a need for some kind of theoretical concept that does at least some of the theoretical work of norms. We suggest, following the article by Suryanarayanan and Kleinman in this volume, that the intellectual space (or the undone social science) occupied by the absence of the term "norms" involves comparative analysis in the broadest sense. Many terms have come to occupy this intellectual space. For example, the early Foucault (1970) used the term "episteme" (what could be viewed as a system of technical norms) to describe continuities across disciplinary differences at a particular time period and likewise differences across those continuities over time, Knorr-Cetina (1999) used the term "epistemic cultures" to describe differences across scientific research fields, and Jasanoff (2004) developed the concept of "civic epistemologies" for her comparative studies of modes of public reasoning and science. It is very likely that the use of the term "institutional logics" in several papers in this volume also emerges in response to comparative work, whether it is across social fields, across time periods, or across forms of public engagement in science. Such comparative work is necessary if one views social science as a nomothetic project that develops generalizations, albeit often limited in scope in the tradition of Merton's middle-range theory, rather than an idiographic project akin to history and ethnography. One could make the argument that all of these social scientists are talking about norms of some sort, but as Hess (1995) suggested, it might be better to think about the diverse approaches in this family as utilizing some variant of the underlying concept of culture in the anthropological sense, that is, as systems of meaning that are both models of and for action. Of course, in contrast to Geertz (1973), who like Merton emphasized uniformity, one must view culture as always a plural field of cultural systems, ideologies broadly defined, or institutional logics that actors continually reformulate, challenge, undermine, and defend. In this sense, a central aspect of culture is the possibility for actors, including collective actors, in subordinate positions in a field or social structure to change it, even if the effects of their actions are not what they originally intended.

FIELDS OF CONTENTION

The discussion up to this point has identified the triangle of structure, agency, and meaning as central to the broad social theory currents in the sociology of science. As the sociologists of scientific knowledge pointed out, the use of social norms and extrafield interests is vexed by complex methodological problems because of the relative autonomy of the field and the tendency for controversies to be settled according to local, technical norms rooted in the standards of subdisciplines and paradigms. (If one does not believe in technical norms at this point, then we challenge the denier to submit an article outside his or her area of methodological competence to a major sociology journal and see what happens.) But we are interested here in a set of research questions that involve a different kind of sociology of science and scientific knowledge than the negotiation of epistemic claims within a research community. Those questions include which research fields become favored for resources and which ones are left to wither away, how what counts as valid science changes in the circulation of knowledge across social fields, and how and why the reward system for scientific research and career attainment has changed over time. Answers to these questions can involve attention to how knowledge is interpreted, shaped, and constructed. As such, the study of the contours of research fields and patterns of done and undone science certainly involves a type of sociology of scientific knowledge. However, there is also a need for explanatory resources that take into account long-term historical changes in the relationship between the scientific field and other social fields. In this sense, there is a need for an analysis of the changing relations of structure, agency, and culture.

Arguably, the most systematic and comprehensive attempt in twentieth-century social theory to bring together the triad in one theoretical framework is the work of Pierre Bourdieu. His background in anthropology and fieldwork in Algeria enabled him to have a semiotic sense of culture, but his ongoing conversation with Marx and Weber also enabled him to situate that understanding in a sociology of action. Whatever one thinks about Bourdieu, it is difficult to accuse him of structural determinism in the sense of a Marxist analysis of science such as Hessen (1971), but his sociology also does not reduce social structure to an epiphenomenal outcome of individual and collective performance.

However, there is a problem in turning to Bourdieu in the sociology of science. He did not write much about science per se, and his sparse writings also reveal significant gaps in his knowledge of the field that to some degree are covered over with polemic. Thus, one's reading of Bourdieu, at least in the sociology of science, needs to be a somewhat recuperative exercise. It does not facilitate matters to find that Bourdieu also changed his views over time. For example, in his mature writings on science (2001), he criticized his own criticisms of Merton as outcomes of his own youthful ambitions and lack of understanding, and he developed a more sympathetic interpretation of Merton based on his biographical background and habitus. Although Bourdieu had earlier followed Mulkay by arguing that Merton presents "little more than a record of the normative rules which are officially professed by its members" (1990, p. 298), in his later work Bourdieu had a more tempered appreciation of Merton, but he also recognized a fundamental difference between his field sociology and Merton's institutional sociology: whereas Merton emphasized the functionality of the reward system and an irenic notion of scientific community, Bourdieu emphasized the scientific field as a site of relations of conflict and cooperation characterized by inequalities of position.

At a more technical level, Bourdieu also expanded on the Mertonian reward system by arguing that scientists accumulated various types of capital, and he developed the convertibility thesis to argue that different types of capital could be translated. For example, a scientist with high symbolic capital (recognition) and cultural capital (knowledge about the field) is in a good position to bring in a grant (financial capital), acquire a new organizational position with more resources (temporal capital), and expand the social network of colleagues (social capital). In turn, as Latour and Woolgar (1986) noted in their extension of Bourdieu's convertibility thesis to their analysis of cycles of credibility, the new levels of financial capital from a grant would enable the scientist to produce a new stream of publications that would lead to new levels of symbolic capital.

Although this example suggests that Bourdieu had influenced some of the early formulations of the sociology of scientific knowledge (see also Knorr, 1977), the sociologists of scientific knowledge did not embrace the concepts of field, capital, and habitus, at least not explicitly, and he became increasingly critical of what he called the "new sociology of science." For example, with respect to Harry Collins's work on interpretive flexibility and the experimenter's regress, Bourdieu writes:

> Remaining within the confines of the laboratory, he does not at all consider the structural conditions of the production of belief, with for example what might be called "lab capital," brought to light by the Mertonians, who showed, for example, as we have seen, that if a discovery is made in a reputed laboratory at a prestigious university it has more chance of being validated than if it emerges in another, less well-regarded one (2001, p. 21).

Whether such criticisms are fair – and against the vast writings of Collins on the gravitational wave controversy, there is much more attention to the positionality of scientists than Bourdieu is willing to admit here – is a question that can be bracketed for the present purposes. Instead, the comment is symptomatic of the tension that emerged between Bourdieu and the leaders of the sociology of scientific knowledge. In a similar vein, he goes on to praise Gilbert and Mulkay (1984) for drawing attention to the different registers of scientific rhetoric (one geared toward the contingencies of the research process and the other toward the impersonal rules of publication), but he disagrees with Knorr-Cetina (1983) over the use of the latter as conscious and, in his terms, "cynical stratagems, oriented toward the glory of researchers" (2001, p. 25). With respect to Latour's Pasteur (1988), Bourdieu writes:

> Rather than seeking the principle of actions where it really lies, in positions and dispositions, Latour can only try to find it in conscious (even cynical) influence and power strategies (thus regressing from a Mertonian collective finalism to a finalism of individual agents) – and the science of science is reduced to the description of alliances and struggles for symbolic "credit" (2001, p. 29).

His perhaps at times unmeasured criticism, or what he reflexively acknowledged as a righteous wrath, especially for actor-network theory, led to predictably hostile reviews of his last writings on science (e.g., Gieryn, 2006; Mialet, 2003; Sismondo, 2005).

Whatever one thinks of Bourdieu's evaluation of the sociology of scientific knowledge (and it was not very detailed), one can separate that issue from the question of the utility of the broad concept of fields in the sociology of science. One of the important contributions of Bourdieu's work methodologically is to situate the question of positions and dispositions at the center of the analytical framework. The social positions can be assessed by examining various levels of capital, but it is the broader relationship among networks of researchers in a field that is of greater interest, that is, the relations of dominant and subordinate networks in a field and the analysis of power that is enabled by the perspective. Although it is possible that some research fields lack such polarities or are characterized merely by a pluralism of camps with relatively equal positions, it is the attention to the polarities of subordinate and dominant positions that is the hallmark of a field sociology of science.

This methodology has many implications, including for research methods. The approach suggests that the social scientist pay attention to the wallflowers of the party, those scientists who are still in the game but scarred by years of shunning and social exclusion. In other words, one would look at the rejected and alternative research programs, including the ones that are still recognized as part of the field and the ones that are pushed out of the field. Whereas scientists at the top of the field either refuse to recognize their domination or view the existence of subordinate networks through the lens of symbolic violence, those at the bottom of the hierarchy often prove to be the best insider sociologists of their own field and its political contours of done and undone science.

Perhaps the most important implication of a notion of fields that is based on relationships of dominant and subordinate networks is the ready possibility of a cultural analysis of the homologies of field positions with distinctions in other social fields. For Bourdieu the habitus, the network of dispositions, was the causal mechanism by which the extrafield distinctions became overlaid on the technical distinctions of the field. To make the rather abstract point concrete with a well-known example, when women were finally allowed into primatology, the theories of human evolution (e.g., the "man the hunter" thesis) altered as did the findings and methods of primatology. However, as Haraway (1989) also showed in the significant last section of *Primate Visions*, there were woman-inflected methodologies that took multiple and unpredictable forms. In effect, when the research field became opened to new entrants (the exogenous shock, to bring in one element of the field sociology of Fligstein & McAdam, 2011), the dispositions of the researchers changed and the configuration of networks of research programs underwent contestation. Although new female entrants to primatology embraced the illusio of the field (the inclination to be part of the field and the ability to participate in it on its own terms), and they also tended to receive patronage support from one of the "big men" of the field, they embraced the illusio in new ways that led to scientific change and, in Harding's terms (1998), also to a condition of stronger objectivity of the science itself.

Thus, one can draw out of field theory a method for analyzing field changes and transitions in the configuration of dominant and subordinate networks of researchers and programs that allows for at least an implicit recognition of interfield relations. As the social composition of the field changes due to new entrants (which can include new social categories based on exogenous statuses, in-migration of scientists from other research fields, and simply new generations of students), intellectual tastes change along with the transitions of habitus. However, Bourdieu's general focus on the relative autonomy of the scientific field and its internal dynamics, like the similar focus of both Merton and the early microsociologies of scientific knowledge, was both a source of great insight and a constraint on theory, especially as the broader field of STS has moved on to study issues of the circulation of scientific knowledge in broader public arenas.

One area for the improvement of field theory involves the tendency to link the habitus to the extrafield biographical trajectories of individuals such as their class culture (Swartz, 1997). But as the previous example suggests, changes in the scientific field are articulated through clashing research programs (or paradigms, or theory-methods-problems packages), and field theory could be improved by having a more explicit formulation of the relations among these research programs. They form a separate cultural field that transects the field of social positions, and the social networks of dominant and subordinate positions of scientists are not necessarily congruent with the intellectual networks of allegiance to dominant and subordinate research programs. In effect, there are two fields at play: a field of social positions based on the capital of individual scientists and their social networks and a cultural field of diacritically related research programs. These research programs are both intellectual forms and objects of intensely passionate support and criticism, a feature of fields that can prevent field analysis from devolving into a kind of rational-actor theory of capital accumulation.

In the political field, there is also a need to think in terms of a double field, but here the clashes of ideas and policy programs appear as ideologies, which are not always completely homologous with the positions of political parties, social movements, trade associations, and other significant collective actors in the field (Hess, 2014). Bourdieu (2005) recognized, at least in passing, this level in his study of the neoliberalization of French housing policy. Although he focused on the habitus of the agents in the housing policy field, he could not have accomplished his analysis without also making reference to the broad clash of political ideologies that motivated the passions of the participants and oriented the alliances and coalitions in the political conflict.

When one seeks to examine the level of the cultural field of contrasting cultural systems that appear in various social fields, the need for a general term emerges. Here, several authors in this volume have opted for the term "institutional logics," borrowed from neo-institutional theories in organization studies. It is possible that the use of the term "logics" may be misunderstood to imply that such systems are necessarily only cognitive and always formally codified, as opposed to the much less restrictive interpretation of "logics" that understands the systems as defined relationally. For example, in the ideological field one understands neoliberalism as what it is not: not social liberalism, not socialism, not localism, and so on. Likewise, it is also possible that against the context of the sociologies of science reviewed here, the term "institutional" will invite misreadings as reversions to Mertonian social norms. Thus, the combination of the two words into "institutional logics" may invite misreadings. But Berman's contrast (this volume) of a science logic and market logic is not about norms that turn scientists into marionettes, just as it is not about well-defined and rationalized systems of thought; rather, the term "institutional logic" refers to a contrastive set of configurations of the discourses, organizations, and everyday practices of knowledge production. This field of contrasting cultural systems also is not an unmoved mover of social action; rather, it is also the object of explicit and implicit contestation. In this sense there is a continuity with field theory, because Berman, like Kinchy and colleagues (this volume), is attending to the issue of which logics come to dominate particular fields of action, and their analysis is sociological because they seek to elucidate the conditions under which some logics become dominant and others marginal.

THE CIRCULATION OF SCIENCE

Perhaps the peace in the feud – the doxa that unites Merton, Bourdieu, and the microsociologies of some of the early laboratory and controversy studies – was the focus on science as a relatively autonomous social field or, in Bourdieu's phrase (2001), as a world apart. But STS as a research field has long ago left the laboratory, and it has followed science and scientists as they circulate through society. Indeed, the focus on the intrafield dynamics of science in Bourdieu was one of the shortcomings in Bourdieusian field theory that led Knorr-Cetina (1981) to the discussion of transcientific fields, just as the idea of following scientists through society became part of the allure of actor-network theory, which moved back-and-forth through the fields of science, politics, industry, and public health (Latour, 1988). The concern with science outside the laboratory is also expressed in the "public turn" of STS, led by Wynne's work on lay publics (1996) and the blossoming of research under the broad rubric of the public understanding of science, as well as in the increasing attention to the issues pioneered by Jasanoff (1990) in studies of science and the state.

Notwithstanding the turn to the study of science "at large" rather than as a "world apart," there is, as yet, no general theory of the interfield dynamics of science. Some elements of an interfield sociology of scientific knowledge in the research programs are outlined above, and there are also elements of an historical and organizational sociology of science and industry in the study of the asymmetric convergence of industrial and academic science (Kleinman & Vallas, 2001), the shifting position of institutional logics in academic science (Berman, 2012), and the shifting valuation of the producer and consumer poles of the scientific field (Albert & McGuire, this volume). Furthermore, these changes are part of the broader historical sociology of advanced capitalism, with the growth of power of multinational corporations, the corporatization of the political field, and the financialization of the economy.

One of the new areas of interfield theory suggested by the articles in this volume might be developed by starting from the different types or species of knowledge, not in the sense of disciplinary differences or even the basic and applied difference, but instead more in parallel with Marx's discussion (1992) of the different species of capital (merchant, industrial, financial, petty) and their inter-relations, and within STS Collins's and Evans's (2007) discussion of types of expertise. From this perspective, an interfield theory of science involves the relations among academic science, including its applied or consumer pole; industrial science and the knowledges embedded in technological standards and innovations; the lay and local knowledge of patients and communities; and occupational expertise. The interest here is not in developing a normative model of how these different types of knowledge should interact; rather, the problem is to develop a sociological analysis of the conditions under which different versions emerge for the circulation of science, other forms of expertise, and scientists across the social fields. We identify two primary types of circulation as an example of one way of moving the theoretical discussion forward.

In the advisory-patronage pathway, scientists provide expertise to identify and solve political and industrial problems, and in turn they receive financial support for their work. As many studies of science policy have documented, scientists work hard to guard their autonomy and also their political neutrality, and when they are successful, their research can provide the epistemic basis, the agreed-upon conditions of reality, for political conflict. However, the authority of scientists to establish the doxa of the political field – the conditions of what is the case upon which debates about what should be done take place – can rapidly break down when their research identifies problems that offend industrial or other political interests. Under those conditions, epistemic rifts emerge, even to the point that scientists are attacked and research programs are defunded (see Hess, this volume). In a similar but less conflictual way, industrial science can emerge to displace the authority of academic science, as Lave (2012) shows in the case of the Rosgen approach to stream restoration. Although the industrial science of Rosgen was a simplification of what academic science had to offer the industry and regulators, it was also preferable because of its consistency with the market requirements of standardized units that could be commodified as financial products in mitigation banking.

The second pathway of circulation, appropriation-scientization, involves the exchange with everyday practices, lay knowledges, and bureaucratic rules and routines. In the original appropriation pattern, the primitive accumulation of modern scientific knowledge, scientific knowledge grew significantly through colonial expeditions, botanical gardens, and museum collections. However, as the scientific field became more esoteric, quantified, and laboratory based, the appropriation of lay and non-Western knowledges became less important in many research fields. Nevertheless, in some cases these other forms of knowledge travel back into science, sometimes through the habitus of new entrants and the diversification of the social composition of science, and sometimes through direct interactions of scientists and publics, as Kinchy and colleagues suggest (this volume). In the scientization side of the cycle, scientific knowledge becomes embedded in everyday practices, lay knowledge (e.g., about health behavior), and bureaucratic routines. As Kinchy (2012) has analyzed, the rule-making regimes of regulatory agencies have also undergone scientization, which can create barriers to participation by social movements. Furthermore, as Moore and Hoffmann (this volume) suggest, in the health field the messaging of government public-health initiatives and the marketing efforts of private-sector health industries lead to transitions in everyday practices that incorporate new knowledges about healthy behavior that are often enveloped in a politics of neoliberal responsibilitization. Likewise, as Schweber (this volume) describes, regulatory policies and voluntary standards also embed scientific rationality and analysis in industrial practices.

Under conditions of epistemic modernization, this second pathway waxes; that is, the scientific field becomes more open to the questions raised by social movements, community organizations, and other sources of lay knowledge, which bring issues of research agendas and undone science to the attention of scientific researchers. However, there is also the countervailing condition of hermeticization, where the scientific field is so esoteric and laboratory based (as in advanced physics) that the trafficking with lay knowledge (except in a few areas, such as cosmology) is highly limited.

These two examples of the circulation of scientific knowledge, the advisory-patronage pathway and the appropriation-scientization pathway, suggest one direction for an interfield theory of science as well as for sociological questions about causal conditions that can be framed from within that theory. In other words, under certain conditions, these cycles can become augmented or weakened. One can study the conditions that facilitate the appearance of epistemic rift in the advisory-patronage cycle or epistemic modernization in the appropriation-scientization cycle. We are in a theoretically uncharted terrain, but one that can build on the studies of classifications of expertise, science-advisor relationships, and lay-expert interactions.

We end here with the beginning, the essays of the volume, and the hope that readers will, like us, find great potential in the essays that have been selected for inclusion, not only because they develop new empirical contributions to the field but also because they help to open up new terrain in the social theory of science, its interfield relations in the contemporary world, and political power.

REFERENCES

Barnes, B., & Dolby, R. G. A. (1970). The scientific ethos: A deviant viewpoint. *Archives Européenes de Sociologie*, *11*, 3–25.

Berman, E. P. (2012). *Creating the market university: How academic science became an economic engine*. Princeton, NJ: Princeton University Press.

Bloor, D. (1991). *Knowledge and social imagery* (2nd ed.). Chicago, IL: University of Chicago Press.

Bloor, D. (1999). Anti-Latour. *Studies in History and Philosophy of Science Part A*, *30*(1), 81–112.

Bourdieu, P. (1975). The specificity of the scientific field and the social conditions of the progress of reason. *Social Scientific Information*, *14*(6), 19–47.

Bourdieu, P. (1981). *The political ontology of Martin Heidegger*. Stanford, CA: Stanford University Press.

Bourdieu, P. (1990). Animadversiones in mertonem. In J. Clark, C. Modgil, & S. Midgil (Eds.), *Robert Merton: Consensus and controversy* (pp. 297–301). New York, NY: Falmer Press.

Bourdieu, P. (2001). *Science of science and reflexivity*. Chicago, IL: University of Chicago Press.

Bourdieu, P. (2005). *The social structures of the economy*. Malden, MA: Polity Press.

Callon, M. (2007). What does it mean to say that economics is performative? In D. MacKenzie, F. Muniesa, & L. Siu (Eds.), *Do economists make markets? On the performativity of economics* (pp. 311–357). Princeton, NJ: Princeton University Press.

Camic, C. (1990). An historical prologue to 'Prolegomena to a theory of social institutions,' by Talcott Parsons. *American Sociological Review*, *55*(3), 313–319.

Camic, C. (2011). Bourdieu's cleft sociology of science. *Minerva*, *49*, 275–293.

Collins, H. (1983). An empirical relativist programme in the sociology of scientific knowledge. In K. Knorr-Cetina & M. Mulkay (Eds.), *Science observed* (pp. 85–113). Thousand Oaks, CA: Sage.

Collins, H., & Evans, P. (2007). *Rethinking expertise*. Chicago, IL: University of Chicago Press.

Daniels, G. (1967). The pure-science ideal and democratic culture. *Science*, *156*, 1699–1705.

Fligstein, N., & McAdam, D. (2011). Toward a general theory of strategic action fields. *Sociological Theory*, *29*, 1–26.

Fligstein, N., & McAdam, D. (2012). *A theory of fields*. Oxford: Oxford University Press.

Foucault, M. (1970). *The order of things*. New York, NY: Vintage.

Foucault, M. (2008). *The birth of biopolitics: Lectures at the Collège de France*. London: Palgrave Macmillan.
Frickel, S., & Moore, K. (Eds.). (2006). *The new political sociology of science: Institutions, networks, power*. Madison, WI: The University of Wisconsin Press.
Geertz, C. (1973). *The interpretation of cultures*. New York, NY: Basic Books.
Gibbons, M., Limoges, C., Nowotny, H., Schwartzman, S., Scott, P., & Trow, M. (1994). *The new production of knowledge: The dynamics of science and research in contemporary societies*. Thousand Oaks, CA: Sage.
Gieryn, T. (2006). Review of science of science and reflexivity. *Contemporary Sociology*, *35*, 185–187.
Gilbert, N., & Mulkay, M. (1984). *Opening Pandora's box: A sociological analysis of scientists' discourse*. Cambridge: Cambridge University Press.
Hagstrom, W. (1965). *The scientific community*. New York, NY: Basic Books.
Haraway, D. (1989). *Primate visions*. New York, NY: Routledge.
Harding, S. (1998). *Is science multicultural? Postcolonialisms, feminisms, and epistemologies*. Bloomington, IN: Indiana University Press.
Hess, D. J. (1995). *Science and technology in a multicultural world*. New York, NY: Columbia University Press.
Hess, D. J. (1997). *Science studies: An advanced introduction*. New York, NY: NYU Press.
Hess, D. J. (2011). *Conference on the political sociology of science and technology studies*. Rensselaer Polytechnic Institute. Retrieved from http://www.davidjhess.org/PSST Conference.pdf
Hess, D. J. (2013). Neoliberalism and the history of STS theory: Toward a reflexive sociology. *Social Epistemology*, *27*(2), 177–193.
Hess, D. J. (2014). Political ideology and the green-energy transition in the United States. In D. Kleinman & K. Moore (Eds.), *Routledge handbook of science and technology studies*. New York, NY: Routledge.
Hessen, B. (1971). *The social and economic roots of Newton's Principia*. New York, NY: Howard Fertig.
Jasanoff, S. (1990). *The fifth branch: Science advisors as policy makers*. Cambridge, MA: Harvard University Press.
Jasanoff, S. (2004). *Designs on nature: Science and democracy in Europe and the United States*. Princeton, NJ: Princeton University Press.
Jones, M. (2009). The rules of the game: Entrepreneurial science. *Social Studies of Science*, *39*(6), 821–851.
Kinchy, A. (2012). *Seeds, science, and struggle: The global politics of transgenic crops*. Cambridge, MA: MIT Press.
Kleinman, D. (2003). *Impure cultures: University biology at the millennium*. Madison, WI: University of Wisconsin Press.
Kleinman, D., & Vallas, S. (2001). Science, capitalism, and the rise of the 'knowledge worker': The changing structure of knowledge production in the United States. *Theory and Society*, *30*(4), 451–492.
Knorr, K. (1977). Producing and reproducing knowledge: Descriptive or constructive? *Social Science Information*, *16*(6), 669–698.
Knorr-Cetina, K. (1981). *The manufacture of knowledge*. New York, NY: Pergamon.
Knorr-Cetina, K. (1983). The ethnographic study of scientific work: Towards a constructivist interpretation of science. In K. Knorr-Cetina & M. Mulkay (Eds.), *Science observed: Perspectives in the social study of science* (pp. 115–140). London: Sage.

Knorr-Cetina, K. (1999). *Epistemic cultures: How the sciences make knowledge*. Cambridge, MA: Harvard University Press.

Kuhn, T. (1970). *The structure of scientific revolutions* (2nd ed.). Chicago, IL: University of Chicago Press.

Latour, B. (1988). *The pasteurization of France*. Cambridge, MA: Harvard University Press.

Latour, B. (1990). Postmodern? No, simply amodern! Steps toward an anthropology of science. *Studies in the History and Philosophy of Science*, *21*(1), 145–171.

Latour, B., & Woolgar, S. (1986). *Laboratory life: The social construction of scientific facts* (2nd ed.). Princeton, NJ: Princeton University Press.

Lave, R. (2012). *Fields and streams: Stream restoration, neoliberalism, and the future of environmental science*. Athens, GA: University of Georgia Press.

Lévi-Strauss, C. (1963). *Totemism*. Boston, MA: Beacon Press.

MacKenzie, D. (1981). Interests, positivism, and history. *Social Studies of Science*, *11*(4), 498–501.

MacKenzie, D. (1983). *Statistics in Britain*. Edinburgh, UK: University of Edinburgh Press.

MacKenzie, D. (1984). Reply to Yearley. *Studies in the History and Philosophy*, *15*(3), 251–259.

MacKenzie, D., & Barnes, B. (1979). Scientific judgment: The Biometry-Mendelism controversy. In B. Barnes & S. Shapin (Eds.), *Natural order* (pp. 191–210). Thousand Oaks, CA: Sage.

Martin, E. (1994). *Flexible bodies*. Boston, MA: Beacon Press.

Marx, K. (1992). *Capital: A critique of political economy* (Vol. 2). London: Penguin Books.

McCarthy, J. D., & Zald, M. N. (1973). *The trend of social movements in America: Professionalization and resource mobilization*. Morristown, NJ: General Learning Press.

Merton, R. (1957). Priorities in scientific discovery: A chapter in the sociology of science. *American Sociological Review*, *22*(6), 635–659.

Merton, R. (1973). *The sociology of science: Theoretical and empirical investigations*. Chicago, IL: University of Chicago.

Mialet, H. (2003). The 'Righteous Wrath' of Pierre Bourdieu. *Social Studies of Science*, *33*(4), 613–621.

Mirowski, P., & Nik-Khah, E. (2007). Markets made flesh: Performativity, and a problem in science studies, augmented with consideration of FCC auctions. In D. MacKenzie, F. Municsa, & L. Siu (Eds.), *Do economics make markets? On the performativity of economics* (pp. 190–224). Princeton, NJ: Princeton University Press.

Mitroff, I. I. (1974). Norms and counternorms in a select group of Apollo Moon scientists: A case study of the ambivalence of scientists. *American Sociological Review*, *39*, 579–595.

Moore, K., Kleinman, D. L., Hess, D., & Frickel, S. (2011). Science and neoliberal globalization: A political sociological approach. *Theory and Society*, *40*(5), 505–532.

Mulkay, M. (1976). Norms and ideology in science. *Social Science Information*, *15*(4/5), 637–656.

Mulkay, M., & Gilbert, N. (1982). Accounting for error: How scientists construct their social world when they account for correct and incorrect belief. *Sociology*, *16*(2), 165–183.

Parsons, T. (1954). *Essays in sociological theory*. New York, NY: Free Press.

Parsons, T. (1990). Prolegomena to a theory of social institutions. *American Sociological Review*, *55*(3), 319–333.

Rapp, R. (2000). *Testing women, testing the fetus: The social impact of amniocentesis in America*. New York, NY: Routledge.

Sahlins, M. (1976). *Culture and practical reason*. Chicago, IL: University of Chicago Press.

Sismondo, S. (2005). Review of *Science of Science and Reflexivity*. *Canadian Journal of Sociology* (Online). Retrieved from http://www.cjsonline.ca/pdf/sciofsci.pdf. (March–April), 1–2.

Slaughter, S., & Rhoades, G. (2004). *Academic capitalism and the new economy: Markets, state, and higher education*. Baltimore, MD: The Johns Hopkins University Press.

Suryanarayanan, S., & Kleinman, D. L. (2013). Be(e) coming experts: The controversy over insecticides in the honey bee colony collapse disorder. *Social Studies of Science*, *43*(2), 215–240.

Swartz, D. (1997). *Culture and power: The sociology of Pierre Bourdieu*. Chicago, IL: University of Chicago Press.

Traweek, S. (1988). *Beamtimes and lifetimes*. Cambridge, MA: Harvard University Press.

Woolgar, S. (1981). Interests and explanation in the social study of science. *Social Studies of Science*, *11*(3), 365–394.

Wynne, B. (1996). Misunderstood misunderstandings: Social identities and public uptake of science. In A. Irwin & B. Wynne (Eds.), *Misunderstanding science? The public reconstruction of science and technology*. Cambridge: Cambridge University Press.

Yearley, S. (2005). *Making sense of science: Understanding the social study of science*. Thousand Oaks, CA: Sage.

PART I
INTERFIELD RELATIONS

UNDERSTANDING CHANGE IN ACADEMIC KNOWLEDGE PRODUCTION IN A NEOLIBERAL ERA☆

Mathieu Albert and Wendy McGuire

ABSTRACT

In this paper, we present and apply a new framework – the Poles of Production for Producers/Poles of Production for Users (PFP/PFU) model – to empirically study how one particular group of academic scientists has responded to neoliberal changes in science policy and funding in Canada. The data we use are from a qualitative case study of 20 basic health scientists affiliated with a research-intensive university in a large Canadian city. We use the PFP/PFU model to explore the symbolic strategies (the vision of scientific quality) and practical strategies (the acquisition of funding and production of knowledge outputs) scientists adopt to maintain or advance their own position of power in the scientific field. We also compare similarities and differences among scientists trained before and after the rise of neoliberal policy. The

☆Authorship in alphabetical order; this paper has been authored equally.

Fields of Knowledge: Science, Politics and Publics in the Neoliberal Age
Political Power and Social Theory, Volume 27, 33–57
Copyright © 2014 by Emerald Group Publishing Limited
All rights of reproduction in any form reserved
ISSN: 0198-8719/doi:10.1108/S0198-871920140000027009

PFP/PFU model allows us to see how these individual strategies cumulatively contribute to the construction of dominant and alternate modes of knowledge production. We argue that the alignments and misalignments between quality vision and practice that scientists in this study experienced reflect the symbolic struggles that are occurring among scientists, and between the scientific and political field, over two competing logics and reward systems (PFP/PFU).

INTRODUCTION

Canada has adopted science policies that promote the use of science and technology to stimulate national economic competitiveness. From the late 1990s to the mid-2000s, the Canadian government increased its investments in research funding and tied these investments to social and economic relevance (Snowdon, 2005; Wolfe, 2002). Existing national funding agencies were restructured and new funding organizations were developed to encourage academic scientists to collaborate with industry, engage in new product development, and generate intellectual property and spin-off companies. By providing incentives for user-oriented research, science policymakers and funders hope that academic scientists will stimulate economic growth by partnering with industry, developing start-up companies that create jobs, attracting investors, and generating revenue for both the private sector and public universities. In short, "good" science is increasingly defined by economic and political stakeholders as "useful" science.

These policy changes appear to promote what Gibbons et al. (1994) have described as a shift from Mode 1 to Mode 2 knowledge production.[1] In Mode 2, academic knowledge is no longer evaluated solely by producers of knowledge through peer-review processes according to traditional academic criteria; it is also evaluated by users, through non-peer-review processes, according to user-oriented criteria. According to a number of scholars, this transformation of knowledge production is associated with the erosion of boundaries between public and private science and an increase in the relative value of applied, problem-oriented research in relation to basic research (Edquist, 1997; Etzkowitz & Leydesdorff, 1998; Slaughter & Leslie, 1997; Ziman, 2000).

In this paper, we explore how basic academic health scientists in Canada responded to user-oriented research funding criteria.[2] While the Mode 2 thesis claims "that the production of knowledge and the process of research were radically transformed" (Nowotny, Scott, & Gibbons, 2003, p. 179), our findings suggest that external pressures are negotiated by individuals in daily practice. Further, we argue that, rather than a sequential and unidirectional shift from Mode 1 to Mode 2, these different regimes of knowledge production coexist in tension, and that the predominance of one over the other is the outcome of struggles among scientists and stakeholders over the legitimate definition of scientific excellence.[3] We present an alternative framework – the Production for Producers/Production for Users (PFP/PFU) model – and illustrate how it can be used to empirically study how individual and groups of scientists negotiate external pressures and construct dominant and subordinate modes of knowledge production through their day to day practices.

POLICY BACKGROUND

In the post-Second World War period up until the 1980s, Canada's science policy focused on broadening and strengthening Canada's knowledge base. From the 1980s onward, the science policy agenda shifted to increasing the utility and use of academic knowledge in order to promote economic competitiveness and innovation. This shift coincided with a broader transition toward a neoliberal political regime in Canada. Policy reforms to promote privatization, deregulation, free trade, regressive taxation, and restrictions on the rights of organized labor were introduced by the federal government throughout the 1980s (Carroll & Little, 2001). In 1988, the first Canadian university research funding program linking academic research with the government's economic strategy, the Networks of Centres of Excellence (NCE) program, was introduced (Fisher, Atkinson-Grosjean, & House, 2001). The NCE program was overseen by Industry Canada and administered through the three main federal granting agencies.[4] As the program evolved, it became increasingly focused on promoting industry-relevant research, university-industry collaboration, and company spin-offs, bringing these criteria to the forefront of funding decisions.

In the 1990s, Canadian politics entered a phase of neoliberal consolidation. Neoliberal policies to support global market expansion dominated policy agendas across policy domains (Carroll & Little, 2001), including science policy. Science and technology moved from the periphery of national politics to take a central role in the federal government's global economic strategy. A cluster of new policy and funding initiatives based on a market logic were implemented to support, facilitate, and accelerate the commercialization of academic research (Berman, 2012; Cameron, 2004; Fisher et al., 2001, 2006; Polster, 2002; Snowdon, 2005; Wolfe, 2002). These included the establishment of the Canadian Foundation for Innovation (CFI) and Genome Canada, increased stable funding for NCEs, and the restructuring of the national health research foundation[5] (Fisher et al., 2001; Wolfe, 2002). New and restructured programs used targeted, industry collaboration or matched funding models which contained new funding criteria and processes (Canadian Foundation for Innovation [CFI], 2008; KPMG, 2002, 2009; Polster, 2002). The effects of these programs are striking. Between 1998 and 2008, the number of patents issued to Canadian universities tripled (from 1,826 to 5,908); the number of licenses rose fourfold (from 788 to 3,343); the royalties generated from licenses tripled (from CAN$15.7 million to CAN$53.2 million); and the number of company spin-offs rose from 454 to 1,242 (Statistics Canada, 1999, 2010). It appears that the transition to a neoliberal political regime is transforming the relationship between science, the state, and the market.

LITERATURE REVIEW

In academic studies of contemporary relationships between science, the state, and the market, the rise in prevalence of university–industry relationships is often interpreted through a lens of stability (i.e., Mode 1 remains dominant) or transformation (i.e., Mode 2 is rising). From a stability perspective, scientists continue to value and are rewarded for producing "good science" as defined by traditional academic criteria (i.e., peer-reviewed grants and publications): they actively resist, minimize, and subordinate market practices according to a traditional Mode 1 academic logic (Albert, 2003; Albert, Laberge, & McGuire, 2012; Gulbrandsen & Langfeldt, 2004; Henkel, 2005; Morris & Rip, 2006). They position their research as user-oriented without changing their practices in order to gain access to material resources from outside the scientific field (Calvert, 2004, 2006; Gulbrandsen & Langfeldt, 2004; Kinchy & Kleinman, 2003). From a transformation perspective, a shift is occurring toward Mode 2, resulting in a new culture and logic of knowledge production in which some scientists carry out user-oriented research to access financial resources or to advance their basic research (Blumenthal, Gluck, Louis, Stoto, & Wise, 1986; Glaser & Bero, 2005; Lee, 2000;

Owen-Smith, 2003; Owen-Smith & Powell, 2003; Slaughter, Archerd, & Campbell, 2004; Tuunainen, 2005). Some research suggests that academic success is increasingly dependent upon scientists' ability to meet user-oriented criteria (Kleinman & Vallas, 2001; Owen-Smith, 2003; Vallas & Kleinman, 2008; Van Looy, Callaert, & Debackere, 2006; Van Looy, Ranga, Callaert, Debackere, & Zimmerman, 2004). Yet even scientists who adopt Mode 2 practices may interpret user-oriented criteria within the framework of traditional science.

This literature provides conflicting evidence as to whether external pressures are changing the way academic knowledge is produced and, if so, the magnitude of change. Even less is known about the impact of a user-oriented vision of academic value developed in the policy field on the evaluation of scientific knowledge within the scientific field. In their review of scholarly reactions to Gibbons et al. (1994) Mode 2 claims, Hessels and van Lente (2008) found little evidence to either confirm or disconfirm changes in academic quality criteria. We do know that Canadian funding bodies now routinely evaluate academic research according to non-academic criteria related to its potential benefits to users through non-peer-reviewed processes, alongside academic criteria and peer-reviewed processes (Fisher et al., 2001; Polster, 2002). Yet these external pressures to incorporate utility as a criterion of scientific excellence do not provide evidence that the distribution of rewards within the scientific community is shifting from a producer to a user-oriented quality vision, inducing scientists to alter their practices (Albert et al., 2012).

THEORETICAL FRAMEWORK

To understand scientists' responses to external pressures in relationship to the scientific reward system, we drew on two concepts developed by one of the authors (Mathieu Albert) in his earlier work: the Pole of PFP and Pole of PFU (Albert, 2003; Albert, Hodges, & Regehr, 2007). These two concepts are inspired by Bourdieu's concept of field (1975, 1993, 1996, 2004) and Gibbons et al. (1994) typology of regimes of knowledge production (i.e., Mode 1/Mode 2). We outline below the key features of Bourdieu's concept of field and describe our reconceptualization of the Mode 1/Mode 2 model into the concepts of Pole of PFP and Pole of PFU.

According to Bourdieu, the social world is composed of multiple social fields in which social agents compete for the acquisition of specific resources (capital). Bourdieu used the concept of field to highlight the relationships (e.g., opposition, alliance, domination, and resistance) between various social agents (individuals, social groups, or institutions) occupying different positions of power in a hierarchically structured social space. In his conceptualization, there are two dominant forms of capital, economic and cultural; all social fields are structured around one of these two forms and are influenced by two other primary forms: social and symbolic capital.[6] Yet each field has a unique logic that differentiates it from other social fields and produces its own specific forms of capital.

The scientific field, which has had a relatively high level of autonomy in the 20th century (at least in Canada) in relation to the market and the political field, has been structured around cultural capital rather than economic capital. In other words, status in science is awarded to those who produce "good science" (a cultural rather than economic product) and the ultimate reward for producing "good science" is peer recognition, which translates into symbolic capital. Scientific capital is the specific form of symbolic capital that structures the scientific field (Bourdieu, 2004). Scientists engage in symbolic struggles over the legitimate definition of scientific excellence as each tries to promote a definition that places value on their own products and scientific practices (e.g., preferred mode of dissemination, chosen paradigms, epistemology, research topic). In doing so, they seek to increase their own chances of accumulating scientific capital and increasing the value of the scientific capital they already hold. Scientists who have accumulated the most scientific capital have the greatest power to influence the activities and distribution of power and resources within the field itself, by holding the authority to say what is worthy and what is not.

While the ultimate currency in this struggle is scientific capital, economic capital is necessary in most disciplines and research areas in order to maintain membership in the scientific field (i.e., to fund research projects, purchase equipment, gain and consolidate social capital through networking during conferences). In the scientific field, the cost of converting economic capital into scientific capital has traditionally been very low, since economic and scientific capital have been awarded according to the same logic: in peer-reviewed grant competitions, scientists with a high level of scientific capital are more successful in gaining access to research funding, which enables them to produce more knowledge, accumulate more scientific capital, and gain access to more grants, in a closed, circular feedback loop. However, neoliberal science policies have introduced a market-based logic into the funding of science. Economic capital is no longer accessible based solely on scientific capital, as more and more grants require scientists to partner with industry, produce business plans, acquire patents and demonstrate the commercial utility of their research. In a previous study, one of the authors (Wendy McGuire) has identified this new form of economic capital as "entrepreneurial capital" which includes industry funding, patents, licenses, new products and technologies, and spin-off companies (McGuire, 2011). Because it is based on a market, rather than a scientific logic, entrepreneurial capital, on its own, has little symbolic value within the scientific field. It can be converted into resources that facilitate the accumulation of scientific capital, but at a higher cost than in the traditional, closed scientific loop. Neoliberal science policy, by producing this new form of economic capital based on a market-based vision of scientific quality, opens up the possibility for scientists to adopt new strategies in the production of scientific knowledge and the struggle for recognition. It also creates the possibility for transformation in the scientific field and, potentially, a shift from a Mode 1 to Mode 2 regime of knowledge production.

According to Gibbons et al. (1994), Mode 1 and Mode 2 are distinct systems of knowledge production that differ in almost every way (Hessels & van Lente, 2008; Pestre, 2003). In Mode 1 science, research problems and agendas are determined within an academic context based on what is most interesting and relevant scientifically; it is organized and carried out within specific disciplines; the individuals involved in research belong to the same research community and the same types of institutions (i.e., academic); the academic community is organized hierarchically; and the quality of academic science is determined by the scientific community through peer review. In contrast, in Mode 2 science, research problems and methods are developed within contexts of application that involve close relationships between scientists and non-scientists; research is carried out by fluid, transdisciplinary teams who bring a heterogeneous set of skills to different stages of research; research is carried out by teams or networks that are loosely institutionalized; power is shared among scientists and non-scientists in less hierarchical ways; and the quality of science is evaluated according to a broad set of criteria, including social accountability, and by a broader range of actors from within and outside the scientific community.

As a typology, Mode 1/Mode 2 has highlighted some of the changes that have taken place at the macrolevel within the scientific system over the last 50 years.[7] As Dominique Pestre pointed out, Gibbons et al.'s book "is highly pertinent as an analytical summary aiming to expose the nature of the transformations under way" (2000, p. 170). However, as the basis for empirical research, we find the model to be of limited use for investigating and understanding how and why changes in knowledge production occur, among particular social actors and within particular social contexts. Because it provides a macro perspective and depicts the shift from Mode 1 to Mode 2 as natural and inevitable (thereby reifying it), this model does not provide the conceptual tools to examine how these two (and any other) knowledge production regimes emerge through social interactions and power struggles. We, therefore, propose an alternative model for studying the social construction of scientific regimes in the neoliberal era. This model emerged from our previous studies (Albert, 2003; Albert et al., 2007) based on the concepts of the Pole of PFP and the Pole of PFU.[8]

The Pole of PFP and the Pole of PFU

The Pole of PFP and the Pole of PFU (see Fig. 1) illustrate alternative visions and active strategies for acquiring symbolic capital among academic scientists. Scientists who mobilize a PFP quality vision (Mode 1 in Gibbon's et al.'s terms) argue that scientific production should, first

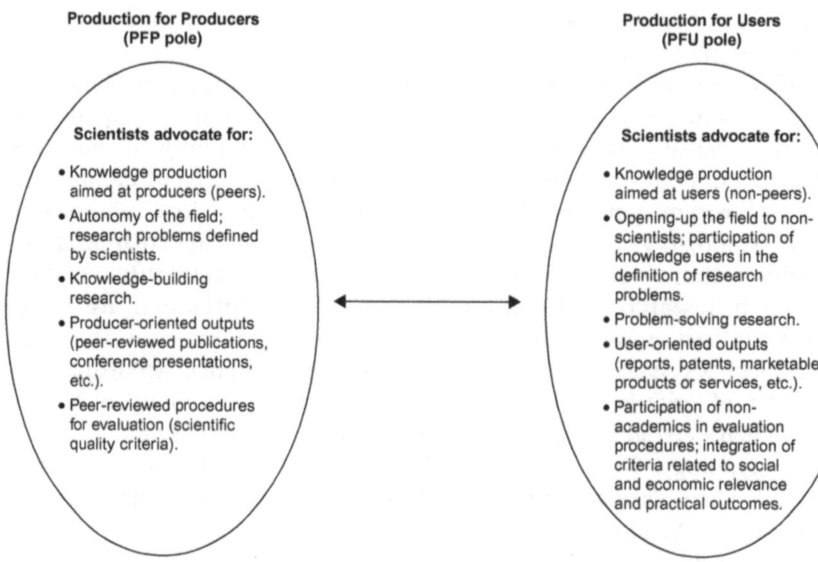

Fig. 1. PFP and PFU Poles.

and foremost, aim for the advancement of knowledge and respond to strict academic criteria of excellence (Albert et al., 2012). They promote production for peers and evaluation by peers as the primary means of acquiring scientific legitimacy as well as the primary mechanism for resource allocation. A high value is attributed to certain types of production, such as articles published in peer-reviewed journals, refereed presentations, and research projects funded by peer-reviewed grants. Using a PFP strategy, these scientists seek to maintain the autonomy of the scientific field in the face of demands arising from external stakeholders, primarily from the economic and political fields. The involvement of non-scientists in the research and evaluation process would likely be perceived as a loss of control of scientists over their own practices. In contrast, scientists who adopt a PFU quality vision and use PFU strategies to accumulate symbolic capital (Mode 2 in Gibbons et al.'s terms) accept utility and socioeconomic relevance as legitimate criteria to evaluate scientific value and advocate for the field to be receptive to external influences (Albert, 2003; Albert et al., 2007). These scientists support knowledge production addressing the problems and needs of the knowledge users. According to a PFU quality vision, studies targeted at issues from outside the scientific community are as legitimate, or more legitimate, than those primarily relevant to academic scientists, and the knowledge outputs that are deemed valuable according to non-academics can also be used to increase legitimacy in the scientific field. The utility of the research, its capacity to find solutions to the real-world problems or to be converted into innovations, can equally constitute a criterion of excellence. Using a PFU strategy, these scientists produce non-academic reports, patents, new products, or technologies. They use the relationships they develop with external organizations as a way of obtaining resources from outside the scientific field and strengthening their position within the field.

Unlike the Mode 1/Mode 2 typology of knowledge production, the two poles in our model do not constitute two mutually exclusive knowledge production regimes into which scientists can be categorized. Rather, these poles have to be seen as ideal types (in the Weberian sense) serving as heuristic tools for comprehending scientists' research practices and definition of legitimate science in the current academic context marked by the autonomy–heteronomy tension. Using data from a study of the responses of basic health scientists at a top-tier Canadian research university, we illustrate how the PFP/PFU model can be used empirically to explore and understand the strategies scientists use to negotiate neoliberal science policies.

METHODS

Using a qualitative case study methodology, 20 basic health scientists affiliated with a research-intensive university in a large Canadian city were interviewed. This university was located in close proximity to a cluster of biotechnology companies, which provided academic scientists with opportunities for industry collaboration and commercialization of academic research. Biotechnology-related research was selected as it is one of the core targets of federal and provincial policies which seek to increase the use of academic research in its broader goals of increasing national economic competitiveness (Biotechnology Council of Ontario [BCO], 2003; Industry Canada, 2001). Scientists in this location conducting this type of research were likely to be subject to significant external pressures and opportunities for commercialization of their research and would likely have had to articulate a response to funding criteria pushing them toward market involvement. Inclusion in the study was restricted to basic scientists involved in the discovery stages of research, excluding applied, clinician scientists and basic scientists involved in clinical trials.

We used a theoretical sampling strategy to generate a sample that had been exposed to two different policy regimes. A theoretical sampling strategy is used to generate data that are useful for testing or generating theory (Patton, 2001). In this case, we wanted to understand whether a change had occurred in scientists quality vision and practice in response to a shift in science policy, indicating a significant change in the scientific field. The year 1992 was used as the cut-off point for the end of the pre-neoliberal policy era and 1999 as the beginning of the post-neoliberal policy era.[9] Established (pre-neoliberal-era) scientists began their academic careers in 1992 or earlier, with 15 or more years' experience, and early stage (post-neoliberal-era) scientists began their academic career in 1999 or later.

Semi-structured interviews were carried out with 11 established and 9 early stage scientists. The interview guide explored four broad areas: (1) the participants' visions of scientific quality; (2) their strategies to acquire research funding; (3) the types of user-oriented criteria they encountered from funding organizations; and (4) the impact of user-oriented funding criteria on their scientific practice and their perceptions of the quality of science overall. Additional data on participants' disciplinary training, work experience, research grants, publications, entrepreneurial practices, and demographics were also gathered from their Curriculum Vitae and a short Background Information Form. Microsoft Word, Excel, and QSR NVIVO were used to manage data and aid data analysis, using several data analysis strategies that are common to qualitative research, including data reduction, constant comparison, exploring contradictions, and testing rival explanations (Miles & Huberman, 1994; Yin, 2003).

RESULTS

To explore participants' visions of scientific quality and their strategies to accumulate capital, we divided scientists' responses toward changes in science funding into two categories: symbolic strategies and practical strategies. Symbolic strategies included the arguments they constructed to defend a particular vision of scientific quality. Practical strategies were the steps they took to accumulate the dominant forms of capital in the field and to convert one form into another when required to gain symbolic capital. Dominant forms of capital were scientific and economic, the former including peer-reviewed publications, tenure, promotion, and awards, and the latter including grants, industry funding, patents, commercializable products, and company spin-offs. Findings are organized in two sections. In the first, we present the symbolic and practical strategies participants used in response to neoliberal science policy to achieve recognition. In the second, the alignment and misalignment of participants' quality vision and practice are examined. In each section, we consider similarities and differences among participants trained before and after the rise of neoliberal science policy.

Symbolic and Practical Strategies to Achieve Recognition

Participants used a variety of symbolic and practical strategies to respond to changes in science policy and defend or improve their position in the scientific field. We found that, both symbolically and practically, the majority of participants leaned toward the PFP pole. As seen in the first column of Table 1 (PFP Pole), 12 out of the 20 participants defended a traditional

Table 1. Traditional and Hybrid Quality Vision and Practice.

	PFP Pole (Traditional)	PFP/PFU Pole (Hybrid)	Total
Quality vision (symbolic strategy)	12	8	20
Accumulation and conversion of capital (practical strategy)	13	7	20

(PFP) quality vision and 13 used a traditional (PFP) strategy of accumulation and conversion of capital. Yet, as seen in the second column (PFP/PFU Pole), a significant number of participants used both PFP *and* PFU strategies, symbolically and practically. Eight of twenty adopted a hybrid (PFP/PFU) quality vision and seven used hybrid (PFP/PFU) strategies of accumulation and conversion of capital.

Participants who argued that traditional (PFP) criteria were the sole legitimate criteria to evaluate scientific excellence ($n = 12$) perceived user-oriented (PFU) criteria and non-peer-reviewed processes as eroding the quality of science without increasing the production of socially and economically relevant innovations. Those who adopted a hybrid quality vision ($n = 8$) perceived user-oriented (PFU) criteria as valid indicators of quality, in addition to PFP criteria, and internalized a hybrid quality vision. The fact that most participants leaned toward the PFP pole was not surprising given that a significant body of research shows that most academic scientists continue to value peer recognition (Albert et al., 2012; Gulbrandsen & Langfeldt, 2004; Henkel, 2005; Hermanowicz, 2009; Hessels & van Lente, 2011).

Our study also confirmed that newer entrants to the field were more likely to view user-oriented (PFU) quality criteria positively (Ambos, Mäkelä, Birkinshaw, & D'Este, 2008; Crowe & Goldberger, 2009; Lee, 2000). Indeed, as seen in Table 2, over half (five of nine) post-neoliberal participants internalized a hybrid (PFP/PFU) quality vision compared with only three of eleven pre-neoliberal participants. Upon first glance, the fact that most pre-neoliberal participants retained a traditional quality vision and most post-neoliberal participants internalized a hybrid vision would seem to indicate movement in the direction toward PFU science.

Yet a different picture emerges when we examine the practical strategies participants used to accumulate economic and scientific capital (see Table 3). Practical strategies included choice of research topics and methods, funding sources, and outputs. Scientists who used traditional (PFP) practical strategies sought out research funding through peer-reviewed, open competition research grants and peer recognition through peer-reviewed publications,

Table 2. Symbolic Strategies (Quality Vision) across Policy Era.

	Traditional (PFP) Quality Vision	Hybrid (PFP/PFU) Quality Vision	Total
Pre-neoliberal	8	3	11
Post-neoliberal	4	5	9

Table 3. Practical Strategies across Policy Era.

	Traditional (PFP) Practice	Hybrid (PFP/PFU) Practice	Total
Pre-neoliberal	5	6	11
Post-neoliberal	8	1	9

promotion, and academic awards. Those who used hybrid (PFP/PFU) strategies did the same, but also sought out industry funding, targeted, or commercialization grants, patented their findings, collaborated with industry and founded their own companies.

We were surprised to find that the quality vision that participants espoused and the practical strategies of accumulation they adopted were often misaligned among both pre- and post-neoliberal scientists. In the next section, we explore alignments and misalignments between quality vision and practice in detail.

Alignments and Misalignments between Quality Vision and Practice

Participants were fairly evenly divided as to whether their quality vision and practice (i.e., strategies of accumulation) were aligned or misaligned (see Table 4). Two groups ($n = 11$) experienced alignment. In the first group, eight participants experienced alignment between their traditional (PFP) vision and practice (row one, column one) and, in the second, three participants experienced alignment between their hybrid vision and practice (row two, column two). A third and fourth group ($n = 9$) experienced misalignment. In the third group (row two, column one), the hybrid (PFP/PFU) quality vision of five participants was misaligned with their traditional (PFP) practices, while in the fourth (row one, column two), the traditional (PFP) quality vision of four participants was misaligned with their hybrid (PFP/PFU) practices.

Traditional (PFP) Quality Vision and Practice: Aligned

An equal number of participants trained in each policy era (four pre- and four post-neoliberal) experienced alignment between their traditional (PFP) quality vision and practice. They perceived user-oriented (PFU) science as a threat to their values or ability to achieve peer recognition. Most had little entrepreneurial experience and little motivation to change their

Table 4. Alignment and Misalignment between Quality Vision and Practice.

	Traditional (PFP) Practice	Hybrid (PFP/PFU) Practice
Traditional (PFP) quality vision	8 (aligned)	4 (misaligned)
Hybrid (PFP/PFU) quality vision	5 (misaligned)	3 (aligned)

practices. For pre-neoliberal participants, a user orientation was almost perceived as a personal affront, as reflected in the following quote:

> I grew up in a tradition where science was something respectable in itself. So when you want to study the universe, you don't talk about applications. You just want to know, right? It's kind of a culture shock for me that everybody, including students, agencies, ask: "What's the use of it? Why are you doing that? What can be cured?" (*Dirk, pre-neoliberal policy era*)

Post-neoliberal era participants responded with greater equanimity toward external pressure for utility but lacked the interest to engage in user-oriented (PFU) practices. According to Damien, "I purposefully insulate myself. I have no interest," while Kate had not yet produced a sufficient body of research that could be translated into commercial products: "I am at the extreme end of not having something I could imagine turning into a patent." They were primarily focused on producing basic knowledge leading to the peer-reviewed publications required to achieve tenure. However, they were aware that they were cutting themselves off from lucrative funding opportunities. Two molecular biologists working in the field of genomics mobilized a traditional (PFP) vision of science to justify their producer-oriented (PFP) practices, arguing that some of their peers who were richly funded through Genome Canada, produced "information" while they produced "knowledge." Collaboration with senior colleagues who fulfilled user-oriented (PFU) criteria was another strategy post-neoliberal participants used to acquire economic resources without having to divert their energy into producing user-oriented (PFU) outputs. Alex, for example, who was a collaborator on a multi-million dollar PFU-oriented grant (Genome Canada), did not engage in any of the commercialization practices required by this funder.

Two pre-neoliberal scientists who adopted a traditional (PFP) quality vision and practice had previously engaged in market activities (PFU practice), but removed themselves due to the cost to their scientific careers. Motivated to develop their basic science research into health improving technologies or treatments, both Robert and Ivan found that the goals of business were at odds with the goals of science. They lost control of their research agendas and became engaged in "technology-type" exercises that were "not science" and, ultimately, failed to produce the desired product. According to Ivan:

> At some point I got so totally frustrated, I just backed out of it. It actually soured me a lot to getting involved in those other things because it took a lot of time. And either to the company or to myself, in terms of academic development and any monetary reward, nothing came out of it. (*Ivan, pre-neoliberal policy era*)

Ivan's investment of time and effort did not result in patents, licenses, or sufficient industry funding to advance his own basic research agenda nor was he able to leverage his industry experience into acquiring user-oriented (PFU) funding. Both Robert and Ivan moved from a position of misalignment back into alignment.

For this group, the cost of engaging in user-oriented (PFU) science was perceived as too high both symbolically, in terms of their academic values, and practically, in terms of their ability to accumulate capital. For the next group, a shift toward user-oriented research funding affirmed their hybrid (PFP/PFU) quality vision as industry-involved basic academic scientists and helped them to enhance their academic reputations.

Hybrid (PFP/PFU) Quality Vision and Practices: Aligned

Only three participants (2 pre- and 1 post-neoliberal policy era) adopted a hybrid quality vision and practice, incorporating both PFU and PFP quality criteria into their quality vision and generating both PFU and PFP outputs. They were successful entrepreneurially, held multiple patents and either collaborated with industry or founded spin-off companies. All three had been able to take advantage of market (PFU) opportunities in order to acquire resources to invest in their traditional (PFP) basic science practices. Pre-neoliberal scientists Sheldon and Jeremy were already engaged with industry and user-oriented (PFU) science before the neoliberal shift in science policy. This policy shift affirmed their (PFU) values which had previously been misaligned with the scientific community. Only one post-neoliberal participant was able to accumulate entrepreneurial capital (e.g., industry funding) and convert it into scientific capital by reinvesting it into his basic science research.

While these scientists were, in principle, in favor of being evaluated according to the use-value of the knowledge they produced, they struggled with user-oriented (PFU) goals and timelines. Yet, overall, for this group, the benefits of PFU science outweighed the risks and they were satisfied with their ability to gain peer recognition. As we will see in the next two sections, the third and fourth groups experienced misalignment between their vision and practices.

Hybrid (PFP/PFU) Quality Vision and Traditional (PFP) Practices: Misaligned

Of the five participants who experienced misalignment between their hybrid (PFP/PFU) quality vision and traditional (PFP) practices, four were post-neoliberal. While they had internalized a user-oriented (PFU) quality vision, these scientists found the costs of accumulating and converting industry funding or commercialization grants into scientific capital too high at this early stage of their career. They had not yet made the big "discoveries" or developed the reputations that would attract business investors and could not afford to divert a portion of their time away from the basic research that would get them the publications needed to acquire tenure. One post-neoliberal participant in this group tried to develop what he saw as a very significant discovery with clear clinical applications but had to abandon these efforts as they took too much time away from his basic research. In Trevor's case, he had moved from alignment between a hybrid (PFP/PFU) vision and practice into misalignment in order to meet the requirements of tenure:

> As a young investigator, I have far more important things to do like publishing papers. However, if I did have to produce patents so that the university granted me tenure, then I probably would pursue it. (*Trevor, post-neoliberal policy era*)

Most post-neoliberal participants, including Trevor, anticipated becoming more involved in hybrid practices after they had achieved tenure. However, as long as traditional (PFP) quality criteria remain dominant in determining tenure, pre-tenure scientists, regardless of their quality vision, are likely to focus on producer-oriented (PFP) practices. Although these scientists experienced misalignment between their hybrid quality vision and PFP practice, this misalignment did not cause them discomfort because it was perceived as temporary.

Only one pre-neoliberal participant who internalized a hybrid quality vision engaged exclusively in traditional (PFP) practices. Jack occupied an unusual position in the field, as a very successful scientist with a high degree of political power. His hybrid quality vision was reflected in his political activities as a senior administrator with a major funding agency, rather than in his scientific activity. Unlike this group, a fourth group of participants whose quality vision and practice were misaligned, experienced a great deal of internal tension.

Traditional (PFP) Quality Vision and Hybrid (PFP/PFU) Practices: Misaligned

All four participants who experienced misalignment between their traditional (PFP) quality vision and their hybrid (PFP/PFU) practices were trained before the rise of neoliberal policy. These scientists adapted their research practices to changes in funding conditions and engaged in patenting, new product development, or started their own companies, in order to access resources from user-oriented (PFU) funding bodies, such as Genome Canada and National Centres of Excellence, and directly from industry. Like pre-neoliberal scientists within the first group we described, who experienced alignment between their traditional (PFP) quality vision and practices, these participants did not perceive user-oriented (PFU) science as "legitimate science." Yet unlike the first group, who perceived the cost of engaging in user-oriented (PFU) science as too high, these participants saw the cost of *not* engaging as too high. Because they did not engage in user-oriented (PFU) practices voluntarily, the dissonance between their PFU practices and their basic science (PFP) values created internal tensions. They managed these tensions by subordinating their user-oriented (PFU) practices to their basic science (PFP) goals, using symbolic strategies to defend their research as "scientific." This can be seen in the following quote:

> It has changed the direction of my research. I have this always in the back of my mind, 'Can we gain intellectual property from this?' Maybe there might have been a more interesting question that we could have delved into. It probably does subtly alter my decision-making. But I would like to think that the primary objective is to do good science. It's not enough to run a successful lab; there isn't enough money out there to do pure research. That's the plain fact of the matter. (*Gerald, pre-neoliberal policy era*)

Despite this misalignment between vision and practice, these participants also spoke of enjoying their involvement in producing products that might have an immediate impact on health. This misalignment was not perceived as entirely negative. As Thomas stated about his work on a new treatment for a highly contagious infectious disease, it "might even be fun up to a point." The greatest satisfaction came from reinvesting the entrepreneurial capital gained from commercialization grants and industry funding into basic science research. As Simon stated, he had only begun to enjoy his scientific career after patenting his findings and acquiring industry funding:

> It was only getting the funding that we did from industry that allowed me to get to that level. It took us up to a whole new level of productivity and quality. We never would have gotten there with granting agencies the standard way. (*Simon, pre-neoliberal policy era*)

The benefits of seeking out and converting entrepreneurial capital into scientific capital appeared, for this group, to have outweighed the costs.

DISCUSSION AND CONCLUSION

In this paper, we have presented and used a new model – the Poles of PFP/PFU model – to empirically study how one particular group of basic scientists has responded to neoliberal changes in science policy and funding in Canada. By using this model, we have been able to understand some of the strategies scientists used to maintain or advance their own position of power in the scientific field, and how these individual strategies cumulatively contribute to the construction of the mode of knowledge production. The PFP/PFU model allowed us to examine various patterns and combinations of producer and user-oriented beliefs and practices. In this study, we were able to see that some scientists participate simultaneously in the construction of a PFP/Mode 1 and a PFU/Mode 2 system of knowledge production, even when this resulted in misalignments between their quality vision and practice.

The use of the PFP/PFU model allowed us to highlight three aspects of basic health academic scientists' experience in the current neoliberal science-policy era. First, our findings show that most pre-neoliberal participants internalized a traditional (PFP) vision of science and most post-neoliberal participants internalized producer and user-oriented (PFP/PFU) quality criteria. However, despite these inter-group differences, the majority of the scientists interviewed stand by the PFP pole both symbolically and practically. These findings echo those of a growing body of work which show the continuing importance of traditional (PFP) quality criteria in academic career success (Gulbrandsen & Langfeldt, 2004; Henkel, 2005; Hermanowicz, 2009; Hessels & van Lente, 2011). While the academic scientific field is not evolving in isolation from the broader social world (see Hess, 2007; Vallas & Kleinman, 2008), it cannot be assumed that change in science policy automatically leads to change in the evaluation and recognition of scientific worth within the scientific community (Bourdieu, 1988, 2004).

Second, we found that a practical strategy is now available to scientists for the accumulation of a new form of economic (entrepreneurial) capital based on PFU criteria and that scientists are able to convert it into symbolic (scientific) capital. Results show that pre- and post-neoliberal scientists have adopted hybrid strategies of accumulation and conversion, suggesting that neoliberal (PFU) science policy has "opened up the space of possibles" within which scientists compete for peer recognition (Bourdieu, 1984, p. 146). However, given the small sample size of this study, it is not possible to identify clear patterns that can explain why particular scientists use PFP or PFU strategies and why some are, while others are not, able to accumulate and leverage their entrepreneurial capital into greater scientific capital. Further empirical research is required to understand the similarities and differences between scientists trained in various science-policy eras (pre-post-neoliberal era) and from various disciplines.

Finally, the PFP/PFU model allowed us to explore alignments and misalignments between quality vision and practice. We argue that these alignments and misalignments reflect the symbolic struggles that are occurring among scientists, and between the scientific and political field, over two competing logics and reward system (PFP/PFU). If a user-oriented (PFU) or Mode 2 system of knowledge production is becoming dominant in the scientific field, and traditional (PFP) strategies no longer yield the same degree of success as they did in the past, we would expect pre-neoliberal scientists to either fight for a return to the previous conditions or to adapt their practices to the new conditions while retaining a traditional quality vision. This was in fact true among the majority of our pre-neoliberal participants. Furthermore, we would expect post-neoliberal era scientists to have internalized a user-oriented (PFU) quality vision and carry out user-oriented (PFU) practices. Instead, we found that most post-neoliberal participants' vision and practice were misaligned, suggesting that a complete transformation in the system of knowledge production (from Mode 1/PFP to Mode 2/PFU) has not occurred.

We cannot draw any conclusions as to which knowledge production system will come to predominate or whether the two will simply coexist in an ongoing state of tension. Struggles over the definition of legitimate science between those scientists clustered around the PFU pole and those clustered around the PFP pole may eventually result in a shift of the power balance and coalesce into a new arrangement between the two structuring forces (PFP/PFU poles) in the field. If that were to occur, the monopoly that traditional (Mode 1/PFP) scientists have held over the definition of scientific quality would be subverted by user-oriented (PFU) scientists and a Mode 2/PFU knowledge production regime would be established, until a new subversion arises. Further research is needed to gain a better understanding of how these symbolic struggles are taking place among other academic scientists and whether they are being structured in similar ways.

In this study, we used the PFP/PFU model to focus on micro level processes, examining how individual scientists construct their careers within an academic environment structured around two opposing poles. In order to understand how these poles come into existence (their social genesis) and how they may alternatively gain predominance, a fruitful approach could be to adopt a meso-level perspective in future research. At the meso-level, our model allows for analysis of group struggles for academic predominance. Scientific groups leaning toward one or the other poles (e.g., clinical scientists vs. basic scientists) may advance their (PFP/PFU) views, practices and evaluation criteria as the most legitimate in order to gain symbolic capital and control the distribution of resources in the scientific field. At both the micro and meso levels, the two poles work as structuring forces as agents shape their actions according to one or the other or both. Specific strategies would depend on the configuration of power in their field and anticipation of potential outcomes. These two poles are not given and their attributes are not permanent, but are constructed by social agents who have an interest in the scientific game (predominantly scientists, but also, and increasingly, governments and business stakeholders) endowed with different powers. Taking on a longitudinal approach would allow us to observe how the field structure developed through time.

Future research should also investigate whether or not those two poles act as structuring forces in fields other than basic health science (we believe they would) and, if so, what is the power configuration between them and how it has come into being. As recent work has shown (Albert, 2003; Kropp & Blok, 2011; Tuunainen, 2005; Ylijoki, 2003), disciplines and research areas are characterized by their own social logic and symbolic order, therefore offering differential responses to funding policy change. Taking into account these specificities, comparative studies would further test the heuristic power of our PFP/PFU model and deepen our understanding of the social logic at work in various scientific fields.

NOTES

1. See also Kleinman and Vallas (2001) and Slaughter and Leslie (1997), for insightful analyses of current changes in knowledge production.

2. The data we used were gathered by one of the authors (Wendy McGuire) during her PhD dissertation research.

3. See Pestre (2000) for a historical perspective on the coexistence of Mode 1 and Mode 2.

4. The Medical Research Council (MRC), the National Science and Engineering Research Council (NSERC), and the Social Science and Humanities Research Council (SSHRC).

5. The Medical Research Council (MRC) became the Canadian Institutes for Health Research (CIHR) with a central mandate to promote knowledge translation Government of Canada (2000).

6. Economic capital is the capital that corresponds to material wealth while cultural capital includes educational or intellectual resources associated with class membership and social status. Social capital refers to the social networks and group memberships that enable access to other forms of capital. Symbolic capital is the form of capital taken by any species of capital when they are perceived as socially legitimate and confer an authority on its holders. In a field where economic capital is dominant, for example, it will also act as a form of symbolic capital, giving social legitimacy to those who possess material wealth (Bourdieu, 1996b, p. 5). Each field has its own specific dominant form of symbolic capital.

7. For insightful and critical accounts of Gibbons et al.'s model see Godin (1998), Hessels and van Lente (2008), Mirowski and Sent (2007), Pestre (2000, 2003), Shinn (2002), and Weingart (1997).

8. These concepts were developed by Albert by building on Bourdieu's concepts of a "subfield of small-scale production" and "subfield of large-scale production" used in his study of the field of French literature (1996a). Similar concepts had previously been developed by Bourdieu in *Le marché des biens symboliques*: "*le champ de production restreinte*" and "*le champ de grande production symbolique*" (1971, pp. 54–55).

9. Before 1992, the NCE program was the only funding program promoting user-oriented processes and products among basic scientists. By the mid-1990s, many new programs had emerged (e.g., CFI, Genome Canada) and user-oriented criteria had been integrated into the three main national funding council decision-making processes. The term post-neoliberal is not meant to suggest that the neoliberal policy era is now over. Scientists that we categorize as "post-neoliberal" were trained during a neoliberal policy era and all scientists continue to practice during this era. Pre and post are used to indicate a shift from traditional to neoliberal science policy.

REFERENCES

Albert, M. (2003). Universities and the market economy: The differential impact on knowledge production in sociology and economics. *Higher Education*, *45*(2), 147–182.

Albert, M., Hodges, B. D., & Regehr, G. (2007). Research in medical education: Balancing service and science. *Advances in Health Sciences Education*, *12*(1), 103–115.

Albert, M., Laberge, S., & McGuire, W. (2012). Criteria for assessing quality in academic research: The views of biomedical scientists, clinical scientists and social scientists. *Higher Education*, *64*(5), 661–676.

Ambos, T. C., Mäkelä, K., Birkinshaw, J., & D'Este, P. (2008). When does university research get commercialized? Creating ambidexterity in research institutions. *Journal of Management Studies*, *45*(8), 1424–1447.
Biotechnology Council of Ontario. (2003). *Getting OnTRAC: Benchmarking Ontario's biotechnology sector*. Prepared for the Biotechnology Council of Ontario by K. Gautam & B. Gordon, BCO.
Berman, E. P. (2012). *Creating the market university: How academic science became an economic engine*. Princeton, NJ: Princeton University Press.
Blumenthal, D., Gluck, M., Louis, K. S., Stoto, M. A., & Wise, D. (1986). University-industry research relationships in biotechnology: Implications for the university. *Science*, *232*(4756), 1361–1366.
Bourdieu, P. (1971). Le marché des biens symboliques. *L'Année sociologique*, *22*, 49–126.
Bourdieu, P. (1975). The specificity of the scientific field and the social conditions of the progress of reason. *Social Science Information*, *6*, 19–47.
Bourdieu, P. (1984 [1979]). In Richard Nice (Trans.), Distinction: A social critique of the judgement of taste. Cambridge, MA: Harvard University Press.
Bourdieu, P. (1988 [1984]). In Peter Collier (Trans.), Homo academicus. Cambridge, UK: Polity Press.
Bourdieu, P. (1993). *Sociology in question*. London: Sage.
Bourdieu, P. (1996a). In Susan Emanuel (Trans.), The rules of art: Genesis and structure of the literary field. Stanford, CA: Stanford University Press.
Bourdieu, P. (1996b [1989]). In Lauretta C. Clough (Trans.), The state nobility: Elite schools in the field of power. Stanford, CA: Stanford University Press.
Bourdieu, P. (2004 [2001]). In Richard Nice (Trans.), Science of science and reflexivity. Cambridge, UK: Polity Press.
Calvert, J. (2004). The idea of 'basic research' in language and practice. *Minerva*, *42*(3), 251–268.
Calvert, J. (2006). What's special about basic research? *Science, Technology and Human Values*, *31*(2), 199–220.
Cameron, D. M. (2004). *Collaborative federalism and post-secondary education: Be careful what you wish for*. Kingston, ON: John Deutsch Institute for the study of economic policy, Queen's University.
Canadian Foundation for Innovation. (2008). *CFI funded projects*. Retrieved from https://www2.innovation.ca/pls/fci/fcienrep.base. Accessed on January 16, 2008.
Carroll, W. K., & Little, W. (2001). Neoliberal transformation and antiglobalization politics in Canada: Transition, consolidation, resistance. *International Journal of Political Economy*, *31*(3), 33–66.
Crowe, J. A., & Goldberger, J. R. (2009). University-industry relationships in colleges of agriculture and life sciences: The role of women faculty. *Rural Sociology*, *74*(4), 498–524.
Edquist, C. (1997). Systems of innovation, *Technologies, Institutions and Organizations*. London: Pinter.
Etzkowitz, H., & Leydesdorff, L. (1998). The endless transition: A triple helix of university–industry–government relations. *Minerva*, *36*(3), 203–208.
Fisher, D., Atkinson-Grosjean, J., & House, D. (2001). Changes in academy/industry/state relations in Canada: The creation and development of the networks of centres of excellence. *Minerva*, *39*(3), 299–325.

Fisher, D., Rubenson, K., Clift, R., Lee, J., MacIvor, M., Meredith, J., … Bernatchez, J. (2006). *Canadian federal policy and post-secondary education: Case report.* New York, NY: Alliance for International Higher Education Policy Studies.

Gibbons, M., Limoges, C., Nowotny, H., Schwartzman, S., Scott, P., & Trow, M. (1994). *The new production of knowledge: The dynamics of science and research in contemporary societies*. Thousand Oaks, CA: Sage.

Glaser, B. E., & Bero, L. A. (2005). Attitudes of academic and clinical researchers toward financial ties in research: A systematic review. *Science and Engineering Ethics*, *11*(4), 553–573.

Godin, B. (1998). Writing performative history: The new new Atlantis? *Social Studies of Science*, *28*(30), 465–483.

Government of Canada. (2000). Bill C-13. An act to establish the Canadian institutes of health research. Government of Canada, Ottawa, ON. Retrieved from http://www.parl.gc.ca/About/Parliament/LegislativeSummaries/bills_ls.asp?lang=E&ls=c13&Parl=36&Ses=2&source=library_prb

Gulbrandsen, M. A., & Langfeldt, L. (2004). In search of 'Mode 2': The nature of knowledge production in Norway. *Minerva*, *42*(3), 237–250.

Henkel, M. (2005). Academic identity and autonomy in a changing policy environment. *Higher Education*, *49*(1–2), 155–176.

Hermanowicz, J. C. (2009). *Lives in science: How institutions affect academic careers?* Chicago, IL: University of Chicago Press.

Hess, D. J. (2007). *Alternatives pathways in science and industry: Activism, innovation, and the environment in an era of globalization*. Cambridge, MA: MIT Press.

Hessels, L. K., & van Lente, H. (2008). Re-thinking new knowledge production: A literature review and a research agenda. *Research Policy*, *28*(4), 740–760.

Hessels, L. K., & van Lente, H. (2011). Practical applications as a source of credibility: A comparison of three fields of Dutch academic chemistry. *Minerva*, *49*(2), 215–240.

Industry Canada. (2001). *Achieving excellence: Investing in people, knowledge and opportunity.* Ottawa, ON: Government of Canada.

Kinchy, A. J., & Kleinman, D. L. (2003). Organizing credibility: Discursive and organizational orthodoxy on the borders of ecology and politics. *Social Studies of Science*, *33*(6), 869–896.

Kleinman, D. L., & Vallas, S. P. (2001). Science, capitalism, and the rise of the 'knowledge worker': The changing structure of knowledge production in the United States. *Theory and Society*, *30*(4), 451–492.

KPMG. (2002). *Evaluation of the networks of centres of excellence: Final report.* Project No. 34725. Prepared by D. Rank. KPMG Consulting LP. Retrieved from http://www.nce-rce.gc.ca/_docs/reports/eval2002.pdf. Accessed on November 20, 2013.

KPMG. (2009). *Evaluation of genome Canada – Final report.* Prepared for Genome Canada. KPMP, Ottawa, ON. Retrieved from http://www.genomecanada.ca/medias/PDF/EN/Five-year_Evaluation.pdf. Accessed on November 20, 2013.

Kropp, K., & Blok, A. (2011). Mode-2 social science knowledge production? The case of Danish sociology between institutional crisis and new welfare stabilizations. *Science and Public Policy*, *38*(3), 213–224.

Lee, Y. S. (2000). The sustainability of university-industry research collaboration: An empirical assessment. *Journal of Technology Transfer*, *25*(2), 111–133.

McGuire, W. L. (2011). *Constructing quality in academic science: How basic scientists respond to Canadian market-oriented science policy – A Bourdieusian analysis*. Ph.D. Dissertation. University of Toronto.

Miles, M. B., & Huberman, M. (1994). *Qualitative data analysis: An expanded sourcebook*. Thousand Oaks, CA: Sage.

Mirowski, P., & Sent, E.-M. (2007). The commercialization of science and the response of STS. In E. J. Hackett, O. Amsterdamska, & M. Lynch (Eds.), *Handbook of science and technology studies* (pp. 635–689). Cambridge, MA: MIT Press.

Morris, N., & Rip, A. (2006). Scientists' coping strategies in an evolving research system: The case of life scientists in the UK. *Science and Public Policy*, *33*(4), 253–263.

Nowotny, H., Scott, P., & Gibbons, M. (2003). Introduction. 'Mode 2' revisited: The new production of knowledge. *Minerva*, *41*(3), 179–194.

Owen-Smith, J. (2003). From separate systems to a hybrid order: Accumulative advantage across public and private science at research one universities. *Research Policy*, *32*(6), 1081–1104.

Owen-Smith, J., & Powell, W. (2003). The expanding role of university patenting in the sciences: Assessing the importance of experience and connectivity. *Research Policy*, *32*(9), 1695–1711.

Patton, M. Q. (2001). *Qualitative research and evaluation methods*. Thousand Oaks, CA: Sage.

Pestre, D. (2000). The production of knowledge between academies and markets: A historical reading of the book 'The New Production of Knowledge'. *Science, Technology & Society*, *5*(2), 169–181.

Pestre, D. (2003). Regimes of knowledge production in society: Toward a more political and social reading. *Minerva*, *41*(3), 245–261.

Polster, C. (2002). A break from the past: Impacts and implications of the Canada foundation for innovation and the Canada research chairs initiative. *The Canadian Review of Sociology and Anthropology*, *39*(3), 275–299.

Shinn, T. (2002). The triple helix and new production of knowledge. *Social Studies of Science*, *32*(4), 599–614.

Slaughter, S., Archerd, C. J., & Campbell, T. I. D. (2004). Boundaries and quandaries: How professors negotiate market relations. *Review of Higher Education*, *28*(1), 129–165.

Slaughter, S., & Leslie, L. L. (1997). *Academic capitalism: Politics, policies and the entrepreneurial university*. Baltimore, MD: Johns Hopkins University Press.

Snowdon, K. (2005). *Without a road map: Government funding and regulation of Canadian universities and colleges*. Research Report W|31Work Network. Canadian Policy Research Networks. Retrieved from http://www.cprn.org/documents/40773_en.pdf. Accessed on November 20, 2013.

Statistics Canada. (1999). *Survey of intellectual property commercialization in the higher education sector, 1998*. Catalogue No. 88F0006XIE – No. 018. Prepared by M. Bordt & C. Read, Government of Canada, Ottawa, ON. Retrieved from http://publications.gc.ca/collections/Collection/Statcan/88F0006X/88F0006XIE2005018.pdf. Accessed on November 20, 2013.

Statistics Canada. (2010). *Survey of intellectual property commercialization in the higher education sector, 2008*. Catalogue No. 88-2220X. Government of Canada, Ottawa, ON. Retrieved from http://www.statcan.gc.ca/pub/88-222-x/88-222-x2010000-eng.pdf. Accessed on November 20, 2013.

Tuunainen, J. (2005). Hybrid practices? Contributions to the debate on the mutation of science and university. *Higher Education*, *50*(2), 275–298.

Vallas, S. P., & Kleinman, D. L. (2008). Contradiction, convergence and the knowledge economy: The confluence of academic and commercial biotechnology. *Socioeconomic Review*, *6*(2), 283–311.

Van Looy, B., Callaert, J., & Debackere, K. (2006). Publication and patent behavior of academic researchers: Conflicting, reinforcing or merely co-existing? *Research Policy*, *35*(4), 596–608.

Van Looy, B., Ranga, M., Callaert, J., Debackere, K., & Zimmerman, E. (2004). Combining entrepreneurial and scientific performance in academia: Towards a compounded and reciprocal Matthew-effect? *Research Policy*, *33*(3), 425–441.

Weingart, P. (1997). From 'finalization' to 'Mode 2': Old wine in new bottles? *Social Science Information*, *36*(4), 591–613.

Wolfe, D. (2002). Innovation policy for the knowledge-based economy: From the red book to the white paper. In G. B. Doern (Ed.), *How Ottawa spends 2002–2003* (pp. 137–156). Toronto: Oxford University Press.

Yin, R. K. (2003). *Case study research, design and methods*. Newbury Park, CA: Sage.

Ylijoki, O.-H. (2003). Entangled in academic capitalism? A case-study on changing ideals and practices of university research. *Higher Education*, *45*(3), 307–335.

Ziman, J. (2000). *Real science: What it is, and what it means*. Cambridge: Cambridge University Press.

NEOLIBERAL CONFLUENCES: THE TURBULENT EVOLUTION OF STREAM MITIGATION BANKING IN THE US

Rebecca Lave

ABSTRACT

Market-based approaches to environmental management are increasingly common. In 1983 when Joeres and David published their pioneering collection, Buying a Better Environment, *the concept was seen as at best novel, and at worst far-fetched. Yet today, conservation and water quality credits are for sale in many developed countries, and the idea of payment for ecosystem services is ubiquitous in environmental policy circles. This paper traces that shift from command-and-control to market-based environmental management through analysis of the evolving practice of stream mitigation banking (SMB) in the US. In the most common form of SMB today, a for-profit company buys land with a damaged stream on it and restores it to produce mitigation credits which can then be purchased by developers to fulfill their permit conditions under the Clean Water Act. Though decidedly noncommercial in origin, SMB was converted into for-profit tradable regulatory*

Fields of Knowledge: Science, Politics and Publics in the Neoliberal Age
Political Power and Social Theory, Volume 27, 59–88
Copyright © 2014 by Emerald Group Publishing Limited
All rights of reproduction in any form reserved
ISSN: 0198-8719/doi:10.1108/S0198-871920140000027010

mechanism in 2000 and has since spread rapidly across the US with the strong support of the US Environmental Protection Agency. Using Bourdieu's field concept as a framework, I argue that the neoliberal transformation of mitigation banking is a product of both relations within the regulatory field, of that field's relations with the fields of science, and of power.

In 1983 when Joeres and David published their pioneering edited volume, *Buying a Better Environment*, market-based approaches to environmental management were little more than a twinkle in economists' eyes. Yet today, there are national and international cap-and-trade markets for carbon dioxide and other pollutants, conservation and water quality credits are for sale in many developed countries, and the idea of payment for ecosystem services is ubiquitous in international environmental policy circles; even the Ecological Society of America advocates market-based techniques for saving the environment. How did putting a price tag on nature come to be viewed as the best way to save it? How has that viewpoint shaped and reshaped environmental regulation and management over the last 30 years? To begin to answer these questions, I focus here on the deep interconnections among federal agency staff, private sector scientists, Executive Branch regulatory reform efforts, and market actors in the neoliberal environmental management practice of stream mitigation banking (SMB) in the US.

Under Section 404 of the Clean Water Act (CWA), anyone who wishes to relocate or culvert an inconveniently located stream is required to make amends for that ecological damage by restoring an equivalent stream elsewhere. Entrepreneurial SMB, a market-based practice for addressing this legal requirement, was rolled out in 2000 and has since spread rapidly, simultaneously creating both a market for ecosystem services and the demand for new metrics to enable that market to function, pulling "science into the service of capital" (Robertson, 2006). In this paper, I first map out the historical shift from command-and-control to market-based management of streams, and then analyze the dynamic relations among mitigation bankers, state and federal regulators, and private sector knowledge producers as SMB policy, practice, and science have evolved since 2000.

The data presented here comes from a collaborative, National Science Foundation-funded project with Martin Doyle, a fluvial geomorphologist at Duke University, and Morgan Robertson, a political ecologist at the

University of Wisconsin. As part of that project, I conducted more than 50 semi-structured interviews with mitigation bankers and restoration designers, and with the agency staff, consultants, and academics who develop the mitigation banking regulations, guidance documents, and metrics that enable the emerging practice of SMB to function. In addition, I analyzed media reports on mitigation banking and government documents on mitigation and market-based environmental management more broadly, focusing on reports, guidelines, and policies released between the late 1950s and 2012.

Based on these data, I argue that the neoliberalization of the environmental regulatory field in the US has been a messy, incomplete process deeply interconnected to parallel transformations in the fields of power and science. Thus while I employ Bourdieu's field concept as an analytic framework to address relations *within* the regulatory field, I also argue that relations *between* fields have been critical to the neoliberal reshaping of stream management in the US, a position that departs from Bourdieu's description of fields as bounded entities. In keeping with arguments in earlier work (Lave, 2012b), by Berman and Hess in this issue, and also by Camic (2011) and Eyal (2013), I argue here for a relational view of fields by demonstrating the central importance of connections among them: transformation in one field (in this case environmental regulation) does not happen in isolation, but is instead deeply tied to changes in other fields (environmental science and politics), and vice-versa.

A RELATIONAL RETHINKING OF BOURDIEU'S FIELD CONCEPT

In Bourdieu's formulation, a *field* is a structured, bounded social arena that provides a specific set of opportunities and constraints to its participants (Bourdieu, 1975, 1983, 1991, 1996a, 1996b, 1998; Bourdieu & Wacquant, 1992; Bourdieu, Wacquant, & Farage, 1994). While Bourdieu presented the state, art, religion, and science as core fields, he also argued that the field concept was useful for analyzing the structure and dynamics of any arena of practice. To grasp the particularities of a field, Bourdieu directs the analyst to focus on the forms of power and prestige (*capital)* valued within it and on the specific ways in which that field shapes its participants' conscious and subconscious practices (the *habitus* it instills). Notably, those participants include anyone whose actions are deeply shaped by the field in question. For example, in the science fiction field, participants would include not just authors and movie directors but also special effects staff, publishing houses, fanzine producers, publicists, reviewers, and even diehard fans.

Internal struggles to define the boundaries and most highly valued forms of capital are one of the central characteristics of Bourdieu's agonistic field concept, and these constant power struggles deeply shape not just the status of individual participants but also the objective structure of the field. This objective structure is organized around an axis running between autonomous and heteronomous poles. At the *autonomous* pole are those participants whose production is controlled most thoroughly by the forms of capital specific to that field; at the *heteronomous* end are those whose production is shaped primarily by outside forces in the field of power. Conflict between the two is one of the central drivers of change within fields.

A field's relative autonomy – the extent to which the field of power shapes the forms of capital most highly valorized – is the second crucial determinant of its overall structure. Relative autonomy can be measured by the extent to which a field, "manages to impose its own norms and sanctions" on its members, including those closest to the heteronomous pole who are, "therefore the most responsive to external demands" (Bourdieu, 1983, p. 321).

While all fields are shaped by their hierarchical relationship to the field of power, Bourdieu downplayed the importance of horizontal interconnections to other fields. Nowhere in his primary theorizations of fields does he describe a field as fundamentally shaped by lateral relations to another field. This is a startling move for a theorist whose default approach is relational. As Gil Eyal writes:

> [W]hile Bourdieu is quite rigorous in applying this relational approach to political programs, works of art, styles of life, or scientific theories, which are all deconstructed into the bundles of relations that constitute them, he does not do the same thing with the distinction between fields themselves .... It is as if the relational approach stopped short before these entities ... and left them in their place as distinct "spheres" whose contents are clearly bounded and well distinguished from one another. (Eyal, 2013, p. 158)

Yet as Camic (2011) points out, as soon as Bourdieu begins to analyze the concrete particularities of an actual field, the importance of interfield relations creeps back in to the analysis. This is very much the case for the field of environmental regulation, which is shaped *both* by its hierarchical relation with the field of power and by its horizontal relations with the field of science. Thus in keeping with Eyal, Camic, and other authors in this

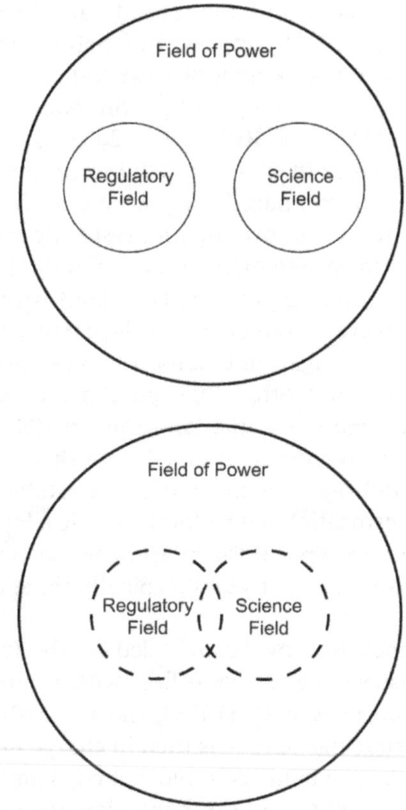

Fig. 1. Shifting to a Relational View of Interfield Dynamics.

issue, I expand Bourdieu's framework to address the ways in which fields can be, and perhaps always are, shaped *in relation* to other fields rather than in the isolation that Bourdieu describes (see Fig. 1).

THE TURBULENT HISTORY OF STEAM MITIGATION BANKING WITHIN THE REGULATORY FIELD

In its current form, mitigation banking is a market-based work-around for a long-standing regulatory issue: how to balance the public goods of development and environmental protection. I describe below how the concept of mitigation banking evolved during the second half of the twentieth century from decidedly noncommercial origins, but before doing so it is worth explaining what mitigation banking looks like today.

Approximately 78% of existing mitigation banks were developed by for-profit companies (Hough & Robertson, 2009, p. 25), typically staffed by people with backgrounds in environmental science, GIS, and land acquisition. For-profit banking companies vary widely in size and capitalization: while some are small and local, owning and restoring only one or two sites, other mitigation firms have regional or even national scope. To create a mitigation bank, these firms acquire rights to land with a degraded stream on it in areas where there is sufficient development pressure that there is likely to be demand for mitigation credits. In some cases mitigation bankers buy land outright; more often they purchase easements, as they are only interested in the stream and the land immediately adjacent to it, and property owners are rarely interested in splitting their property by selling a slice out of it. Once land rights of some form are established, the mitigation banking firm hires restoration consultants to develop both a design to restore the stream and a proposal for monitoring particular aspects of that design over a specified number of years (typically three to five) after project completion.

The state Interagency Review Team – led by the local Army Corps of Engineers (Corps) district regulatory office, with representatives from the Environmental Protection Agency (EPA), the US Fish & Wildlife Service (FWS), and relevant state agencies – is then in charge of reviewing the bank proposal and eventually the bank as implemented, signing off on the release of stream mitigation "credits" as the bank demonstrates that it has met particular success criteria over time. Once certified by the Interagency Review Team, released mitigation credits can be purchased at a negotiated price by anyone needing a permit to offset the ecological damage caused by their impairment of a stream somewhere within the "service area" of the bank: the watersheds within which restoration at the bank site is viewed as compensating for ecological harm.

SMB and its older cousin, wetlands mitigation banking, are one of the oldest continuously functioning ecosystem service markets in the world, and they are promoted as poster children for neoliberal environmental management: win-win exemplars of the use of markets to improve the efficiency of environmental protection. Yet it is critical to realize that, as with most existing ecosystem service markets, mitigation banking was created not by the unconstrained action of the free market or by the voluntary actions of environmentally minded individuals, but through governmental intervention. Mitigation banking was first created and then substantially strengthened through the actions of federal agencies, with the strong encouragement of Executive Branch regulatory reform efforts under Carter, Reagan, Clinton, and G. H. W. Bush. Nor does government intervention end once the market is established. As Robertson notes, the Corps both certifies the credits produced by mitigation banks and requires permit applicants to purchase them (Robertson, 2004, p. 103). Mitigation banking is thus a product less of the invisible hand of the market than of the mailed fist of the state.

In this section I describe the internal struggles within the regulatory field that produced mitigation in the late 1950s, and converted it into a tool of market-based environmental management in the 1990s.

A Brief History of Mitigation

> The word mitigation is derived from the Latin verb *mitigare*, which, significantly, can mean both "to make less severe," or "to appease, assuage or pacify." Accordingly, in the context of the Section 404 program, mitigation entails actions taken to make permitted impacts to the aquatic ecosystem less severe. Observers of the history of mitigation, however, can be forgiven for concluding that mitigation also entails the appeasement or pacification of many divergent interests. (Hough & Robertson, 2009, pp. 15–16)

The concept of mitigation was first applied to the environment in US law in the amendments of the 1958 Fish and Wildlife Coordination Act (FWCA). In a move that predates the landmark environmental legislation of the early 1970s by more than a decade, the FWCA argued for the equivalence of environment and development as public goods:

> For the purposes of recognizing the vital contribution of our wildlife resources to the Nation, the increasing public interest and significance thereof due to expansion of our national economy and other factors, and to provide that *wildlife conservation shall receive equal consideration and be coordinated with other features of water-resource development programs* through the effectual and harmonious planning, development, maintenance, and coordination of wildlife conservation and rehabilitation ..., the Secretary of the Interior is authorized ... to provide assistance to, and cooperate with, Federal, State, and public or private agencies and organizations ... in carrying out ... measures necessary to effectuate the purposes of this Act .... (US Congress, 1958, italics added)

Note the total absence of regulatory teeth: the Secretary of the Interior is not instructed to regulate or enforce, but to assist and cooperate. Thus while the FWCA served as, "the ticket which admitted the State and Federal wildlife agencies into the water resources game" (Greenwalt, 1979, p. 79), that ticket only took them so far. The FWCA effectively placed, "the FWS and other commenting agencies in the position of mitigation negotiators that lack regulatory authority" (Short, 1988, p. 30). FWS staffers reported great difficulties in convincing what they commonly referred to as the "construction agencies," such as the Corps and state departments of transportation, to comply with the spirit, much less the letter of the FWCA (e.g., the rampant cynicism at the 1979 Mitigation Symposium, and Short, 1988).

FWS and state wildlife agency staff's primary resource in this highly unequal contest was language in the FWCA that called on them to, "describe the damage to wildlife attributable to the project and the measures proposed for mitigating or compensating for these damages" (US Congress, 1958, Section 2(b)). Mitigation became a central tactic for salvaging *something* in the frequent cases where FWS staff could not prevent damage to water resources, originating as a weapon of the weak rather than an instrument of market triumphalism. And as Scott originally noted, weapons of the weak are not always as effective as their wielders might hope (Scott, 1985). As reported in Horak (1979) and Redmond (1992), many mitigation requests were rejected outright, a substantial portion of those accepted were never implemented, and many of those that were implemented were partially or entirely unsuccessful in meeting their environmental goals.

In the early 1970s the mitigation policy landscape went through what initially appeared to be a dramatic shift. The federal courts (particularly the Fifth Circuit Court of Appeals in Zabel v. Tabb (430 F. 2nd 199 (5th Cir 1970))) and the passage of the National Environmental Protection Act in 1970 changed water resources mitigation from a request to a requirement but gave no guidance on the extent of mitigation required, or even how to define it in the first place. Initially this lack of definition appeared not to be a problem because revisions to the Federal Water Pollution Control Act (later retitled the Clean Water Act) in 1972 seemed to give FWS powerful ammunition and allies to prevent impacts to water resources in the first place. Specifically, Section 404 of the CWA regulates the discharge of dredged or other materials into waters of the US with the intent of minimizing allowable damage to aquatic ecosystems. Mitigation was not even mentioned in the CWA as a technique for compliance with Section 404 because legislators assumed that EPA and the Corps (to whom they gave joint responsibility to administer the program) would use their authority to modify or deny permits to *prevent* harm to aquatic systems rather than allowing harm and then asking permit applicants to make amends for it through mitigation. In practice, however, the Corps proved unwilling to deny permits except in the most egregious cases: for example, only 0.25% of all applications were denied from 2004 to 2005 (Hough & Robertson, 2009, p. 27), while EPA has almost entirely refused to exercise its veto power over Corps permits, overturning just 11 permits in the first 36 years of the program's existence (*ibid.*, p. 22). As Hough and Robertson described it, the Corps had hoped that EPA would, "wear the black hat" so that it could, "extract concessions from the permittee without wielding overbearing regulatory force" (*ibid.*, p. 21), but EPA seemed to think it looked better in white.

The EPA was then stuck in an analogous position to FWS. The CWA mandated protection of aquatic resources. If the agency was unwilling to accept the political costs of preventing harm, there had to be a mechanism through which permit applicants could offset that harm. Mitigation thus grew into a central component of CWA regulation and implementation, with EPA piggybacking on what was by then nearly two decades of work by the FWS, supporting it in a struggle with recalcitrant Corps staff that was only resolved in 1990 after multiple court cases and years of acrimony (see Hough & Robertson, 2009 for a useful summary).

Eventually, the Corps and EPA agreed to regulatory guidance for mitigation that consists of three stages that are supposed to be performed in sequence: *avoidance* of impacts to the extent possible, *minimization* of any impacts deemed unavoidable, and *compensation* for unavoidable impacts via ecological restoration. The EPA's and the Corps' continued gravitation to the path of least political flak converted what was intended to be the third stage of the mitigation process, compensation, into the first and final step in the majority of cases, as it is often more palatable for regulatory agencies to ask developers and public works agencies to pay compensation than to thoughtfully redesign proposed projects. The end result, in Hough and Robertson's words, is that, "Compensatory mitigation is so central to discussions of mitigation that 'compensation' is often mistakenly held to be synonymous with 'mitigation', even among the most experienced observers of the [Section 404] program" (*ibid.*, p. 23). Quoting Houck, they argue that compensatory mitigation is, "'the most seductive concept in the field of wetlands protection' ... because of the temptation to resolve tough permit decisions by seeking more aggressive compensation packages from applicants rather than by fully exploring avoidance and minimization" (Houck, 1989, p. 836, cited in Hough & Robertson, 2009, p. 23).

Thus the practical and political considerations of implementation have transformed the CWA over time from a focus on prevention to a program of indulgences centered on compensating for harms committed. And once compensation was firmly in place as a mitigation strategy in the environmental regulatory field, it was not a large leap to imagine marketizing the whole process. Rather than invent a new structure to do so, proponents of neoliberal regulatory reform latched onto an existing and firmly nonentrepreneurial precedent: the mitigation bank.

Mitigation Banking Begins

While the Corps and EPA struggled over mitigation, the FWS and its allies at the National Marine Fisheries Service and the state wildlife agencies found themselves more or less alone facing down the political pressures EPA and the Corps were so busy avoiding, tarred as antidevelopment and obstructionist for trying to carry out their wildlife protection mandate under FWCA (Short, 1988, p. 30). In an attempt to make their mitigation recommendations more palatable, FWS and agencies in a few states, most notably California, Oregon, and North Dakota, began to experiment with "mitigation banks," 13 of which were up and running by the early 1980s (Short, 1988). This initial, non-entrepreneurial form of mitigation banking was modeled not on commodity markets but on savings accounts:

> Mitigation banking has been likened to maintaining a bank account. A developer implements measures to create, improve, or preserve fish and wildlife habitat prior to an anticipated need for mitigation for project impacts. The benefits of these measures are quantified as mitigation credits for the developer and placed in a mitigation bank account from which withdrawals can be made. When the developer proposes a project which will result in unavoidable losses of fish and wildlife resources, the losses are quantified as debits using the same method that was used to determine bank credits. A withdrawal equal to that amount is deducted from the bank balance. The debiting process can be repeated as long as mitigation credits are available in the bank. (Soileau et al., 1985, quoted in Short, 1988, p. iii)

Early mitigation banks were both noncommercial (credits were not for sale) and not for profit. Instead of private sector companies, 12 of the 13 initial banks were developed by ports and state Departments of Transportation for their own future permitting convenience. These initial mitigation banks were notable for their avoidance of all things commercial; in FWS' 1988 report on mitigation banking (Short, 1988), for example, there are multiple mentions of the need to avoid the appearance that permits are being bought or sold, and the single mention of markets is in relation to land costs, not credit sales. The first commercial sale of wetlands credits did not occur until 1986, and even then it was as an afterthought at a bank intended primarily for in-house credits for the only private sector banker at the time, a major oil company. The first truly entrepreneurial bank began selling credits in Florida in early 1994 (Hough & Robertson, 2009, pp. 24–25). As late as 1992, economists still felt it necessary to call the federal agencies on the carpet for avoiding the commercial language that described what banks were about to become: commodity markets (e.g., King, 1992).

THE MARKETIZATION OF MITIGATION BANKING: RELATIONS BETWEEN THE REGULATORY FIELD AND THE FIELDS OF POWER AND OF SCIENCE

What transformed the character of mitigation banks from the upright financial prudence of savings accounts to the speculative entrepreneurialism of commodity markets? I argue here that the key drivers arose from relations between the environmental regulatory field and two other fields: politics and science (see Fig. 2). Executive Branch regulatory reform efforts beginning in the late 1970s under the Carter Administration and stretching into the 1990s under the Clinton Administration pushed the neoliberalization of many

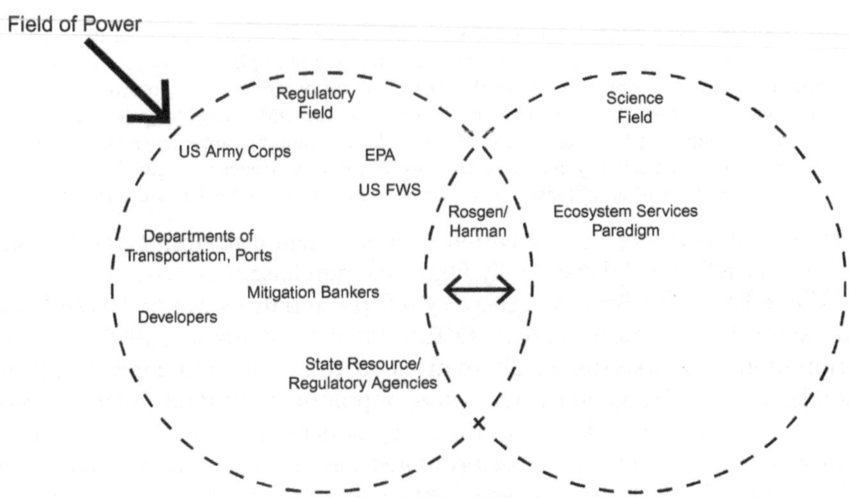

Fig. 2. A Relational View of Interfield Dynamics.

policy arenas, including environmental regulation. This pressure from above found logistical and ideological support in a parallel transformation of the environmental science field with the rise of the ecosystem services paradigm in ecology and ecological economics. Together these interfield relations laid the groundwork for entrepreneurial mitigation banking.

The Executive Branch Embraces Regulatory Reform Again, and Again, and Again[1]

Executive Branch attempts to retool regulation to neoliberal specifications began in 1978 and received strong backing under successive Democratic and Republican administrations (see Robertson's, 2013a, 2013b, blogposts at http://wetlandia.blogspot.com/2013/08/jimmy-carter-and-archaeology-of-market.html). While deregulation had some legs, particularly under Reagan, attempts to reregulate along neoliberal lines had surprisingly little impact until the mid-1990s. Clinton's regulatory review taskforce commented ruefully in a 1993 report that, "Despite more than a decade of high-level support, innovative approaches are not widely used by agencies" (National Performance Review, 1993). And writing in a Cato Institute magazine in 1997, Weidenbaum decried the long-standing ineffectiveness of Executive Branch regulatory reform efforts, complaining that agency staff had become skilled constructors of Potemkin cost/benefit analyses:

> [E]xperienced bureaucrats are accustomed to offering lip service to the various circulars issued by the OMB [Office of Management and Budget] to implement presidential policies. The regulators sincerely believe that they have an overriding duty to protect their programs from the "green eyeshade economists" at the OMB and from the political pressures emanating from the White House. The resultant response, a ritual performance of some perfunctory economic analysis, enables the agencies to ignore the spirit of the entire effort while still meeting the formal requirements. (Weidenbaum, 1997)

Yet as the existence of mitigation banking demonstrates, market-based regulatory reforms did eventually find some purchase.[2]

While Executive Branch regulatory reform initiatives can be traced back to the Ford Administration (e.g., OIRA, 1998; Weidenbaum, 1997), explicit promotion of market-based alternatives to command-and-control regulation began with President Carter, whose appointment of Paul Volcker seems not to have been the fluke some critical scholars believe (e.g., Harvey, 2005). Carter's Executive Order 12044 promoted neoliberal regulatory reform in order "to achieve legislative goals effectively and efficiently ..., [without imposing] unnecessary burdens on the economy" (Carter, 1978). Carter's

Regulatory Council, for example, was charged with developing a list of innovative techniques and recommendations to: "find ways to achieve their goals with reduced burden on the private sector … [while also promoting] innovation, putting private ingenuity to work, [and] finding better long-term solutions to regulatory problems" (Carter, 1980). The Council's Project on Alternative Regulatory Approaches promoted eight primary approaches, most of which have clear ties to neoliberal philosophies. As described in a 1980 report, the list of alternative approaches began with:

- Marketable Rights: Distributing a limited number of rights to scarce resources that private parties can then buy, sell, or trade as market needs dictate. This can remove the government from difficult, contentious, and lengthy decisions about who can "best" use the limited resources.
- Economic Incentives: Using fees or subsidies (rather than government-enforced standards) to encourage private sector achievement of regulatory goals. This approach removes the government from having to eliminate or directly restrict unwanted activity, but creates an incentive for the private sector to limit the activity itself. (US Regulatory Council, 1980, p. v)

The remaining alternative approaches included Performance Standards to increase flexibility by replacing regulations that spell out particular paths to compliance; Compliance Reform to supplement government enforcement with "market-oriented mechanisms including third-party compliance monitoring … and supervised self-certification"; Enhanced Competition; Information Disclosure to "replace centralized government decisions with informed freedom of choice among many users," and Voluntary Standards (op. cit.). Yet while this and other Carter administration initiatives talked the neoliberal talk, very little neoliberal walking seems to have occurred at the time.

Reagan picked up where Carter left off, making regulatory reform (which he unsubtly restyled as "regulatory relief") one of the "four pillars" of his initial platform and encouraging the federal regulatory agencies to embrace neoliberal practices. Immediately after he took office, Reagan took the notable step of requiring agencies to demonstrate that the economic benefits of any major proposed regulation exceeded its costs (Reagan, 1981). He established a high-powered group – the Taskforce on Regulatory Relief – chaired by Vice-President Bush, to operationalize the cost/benefit analysis of federal regulation (OIRA, 1998; Weidenbaum, 1997). Section 404 of the CWA, one of the primary drivers of mitigation, was an explicit target of the Taskforce's effort, which at one point empowered the Corps to ignore EPA's guidance documents and skip the avoidance and minimization stages of permit review in order to move straight to mitigation (Glubiak, Nowka, & Mitsch, 1986, p. 146), although this was later overturned through court actions (Hough & Robertson, 2009, pp. 20–21).

By his second term, Reagan had deemphasized regulatory reform (Weidenbaum, 1997), but President G. H. W. Bush took up the fight again, replacing the Task Force on Regulatory Reform with the White House Council on Competitiveness in 1989, and putting his own vice-president, Dan Quayle, in charge. In 1991, Bush gave the Council additional powers to directly revise regulation when mere review did not have the transformative impact he had hoped (Hilts, 1991). Section 404 of the CWA was once again a target, as the Council published new, less restrictive guidelines for determining what constituted a wetland, and promoted mitigation banking and other entrepreneurial approaches to environmental protection (Hough & Robertson, 2009, p. 28). The Council's expansion of Executive Branch power caused a great deal of controversy, particularly given its habit of allowing corporations to lobby without disclosure (Weidenbaum, 1993). Thus when Clinton took office in 1993, one of his first actions was to insist on transparency for his own neoliberal regulatory reform efforts.

In other ways, though, Clinton was very much in line with his predecessors, using an early Executive Order (Clinton, 1993) to establish a regulatory review council, the National Performance Review, led by his Vice-President Al Gore. The National Performance Review was tasked with increasing flexibility, reducing regulatory burden, and reducing costs of compliance. It released a number of reports promoting neoliberal reforms, including the 1995 *Reinventing Environmental Regulations*, which argued that, "[m]arket incentives should be used to achieve environmental goals whenever appropriate," and stated that while command-and-control regulation may remain a policy option,

> [W]e must expand available policy tools to include new and innovative ways to achieve greater levels of environmental protection at a lower cost. For example, we have learned that setting "performance standards" and allowing the regulated community to find the best way to meet them can get results cheaper and quicker – and cleaner – than mandating design standards or specific technologies. We can promote both lower-cost environmental protection and innovation in pollution control and prevention technology. Using performance standards along with economic incentives encourages innovation. The lowest-cost and most effective strategies earn a greater return in the marketplace. (National Performance Review, 1995)

In keeping with statements such as the one above and with National Performance Review reports recommending markets for air and water pollution, the Clinton administration strongly promoted entrepreneurial mitigation banking. And it appears that they succeeded in neoliberal re-regulation where Carter and Reagan did not: the market in wetland mitigation banking credits was up and running by the mid-1990s, and entrepreneurial SMB began in 2000 with sales driven by Corps' permit requirements for compensatory mitigation.

Yet it is not enough for governments to promote neoliberal regulatory reforms such as the creation of ecosystem service markets, or even to mandate purchases within those markets. In order for any market to function it must have a commodity to trade, one that is defined by simple, reliable metrics. As I will argue below, this fundamental need for commodity definitions deeply implicates environmental science in neoliberal environmental management (Robertson, 2006). In the case of SMB, the first step was conceiving of ecosystems as a set of services that could take on measurable economic values: the ecosystem service paradigm.

The Ecosystem Service Paradigm in the Field of Science

The concept of ecosystem services developed within academia during the same time period as the regulatory reform efforts described above (roughly the late 1970s through mid-1990s), and "ascended boldly into environmental policy discussions" (Dempsey & Robertson, 2012, p. 758) as the basis of market-based environmental approaches starting in the late 1990s (Dempsey & Robertson, 2012; Gomez-Baggethun, de Groot, Lomas, & Montes, 2010; Gomez-Baggethun & Ruiz-Perez, 2012). As Gomez-Baggethun et al. describe it, "use of the ecosystem services concept has transcended the academic arena to reach governmental policy as well as the non-profit, private and financial sectors" (2010, p. 1209). Like mitigation banking (and unlike federal regulatory reform efforts), the ecosystem service paradigm [ES] was not conceived with neoliberal intent. Instead, it originated as, "an eye-opening metaphor intended to awaken society to think more deeply about the importance of nature and its destruction," and the capitalistic, monetized language of ecosystem services was seen by its instigators as distasteful but "necessary to awaken a public deeply embedded in a global economy and distant from natural process" (Norgaard, 2010, p. 1219). To critical social scientists, the naiveté inherent in assuming that the language of monetization and capitalism can be firewalled off from the institutions and practices that embody them is striking. In Dempsey and Robertson's words, "the belief that money's use as universal equivalent can be separated from the operation of capital to generate a profit – a belief common among both scientists and policy-makers in ES, … [is] something that critical researchers must regard as a key blindness" (2012, p. 772).

The origin of ES is typically traced back to key work by the Odum brothers and the Ehrlichs, among others (Ehrlich & Ehrlich, 1981; Helliwell, 1969; King, 1966; Odum & Odum, 1972) who sought to use capitalistic metaphors to raise environmental awareness. By speaking the lingua franca of late twentieth-century life – monetary value – early ES proponents hoped to galvanize those to whom the language of environmentalism seemed to be unintelligible. From this initial work by ecologists and ecological economists, ES began to spread into international environmental policy circles via a cameo appearance in the Convention on Biological Diversity in 1992.

In 1997, ES made the leap into environmental policy ubiquity through the publication of two germinal pieces: an ecological economics paper in *Nature* that put a price tag of $33 trillion on the entire range of global ESs (Costanza et al., 1997), and ecologist Gretchen Daily's edited volume *Nature's Services* (1997). As Dempsey and Robertson note, "[t]ogether these publications carried the banner for a movement within policy and economics that has reached the highest levels of global environmental governance and development policy" (2012, pp. 758–759). These texts sparked the development of a wave of assessment techniques to enable valuation of the broad range of ESs; the drive to save nature by incorporating it into capitalism was on. By the time the Millennium Ecosystem Assessment was released in 2005, ES had become a central framework of international environmental policy.

Although in theory valuation and commodification (putting a price tag on nature vs. actually selling it in a market) are separable, in practice, valuation provides an absolutely critical first step, paving the way for transforming ecosystems into commodities. As Gomez-Baggethun and Ruiz-Perez note, context matters, and our current context is broadly neoliberal:

> Since valuation outcomes are inevitably attached to the ideological and institutional structures in which environmental policy operates, valuation practitioners cannot expect monetary figures to be neutral in relation to the process of nature commodification … [Thus the] economic framing of the environment and monetary valuation methods cannot be considered neutral tools. (2012, p. 624)

ES' dependence on the language of capitalism made it a very smooth fit with the types of neoliberal governance reform efforts described above.

Originally conceived as a tool to galvanize environmental awareness, ES morphed instead into a central tool in both justifying and implementing market-based environmental management efforts such as SMB.

THE EVOLUTION OF ENTREPRENEURIAL STEAM MITIGATION BANKING, 2000–2008

The transformation of mitigation banking into an instrument of neoliberal regulatory reform was importantly shaped by interfield relations between the field of environmental regulation and the fields of power and of science. In this section, I address how internal struggles within the regulatory field shaped the development of entrepreneurial mitigation banking from 2000 to 2008.

Once SMB was reconfigured into a tool of market-based regulatory reform, it expanded very quickly: the first entrepreneurial stream mitigation bank was established in 2000, and by 2011 there were 240 in the Corps' registry (Doyle & Shields, 2012, p. 2). Even with an existing regulatory form, political support from the field of power, and ideological and practical support from the ES paradigm in the field of science, getting this market in stream credits up and running was not simple. Establishing a stream credit as a standardized, tradable commodity depended on creating equivalence between the reach of stream to be damaged by a permit applicant and a reach of stream previously restored by a mitigation banker. But what sort of commodity can be made out of a stream? The customary ways we define commodities – by weight, volume, or number – were clearly inapplicable. Instead regulators and bankers needed new metrics capable of paring away the complexity and ecological specificity of both the damaged and restored reaches, abstracting them into easily measurable, comparable units (Cronon, 1991). The need to create these kinds of metrics strongly connected the environmental science and regulatory fields.

In entrepreneurial wetlands mitigation banking, which got its first tentative start in the mid-1980s, the development of these metrics has been the work of decades of scientific labor, primarily on the part of university- and agency-based scientists. In sharp contrast, when SMB began in 2000 there was a widely used metric ready to hand: the stream classification system developed by consultant David Rosgen (Lave, 2012a, 2012b; Lave, Doyle, & Robertson, 2010). The Rosgen classification system focuses on channel form, using the transport of water and sediment as a proxy for the improvements in water quality and aquatic ecology that are the primary drivers of stream mitigation. Despite the classification system's physical focus, and despite the fact that Rosgen himself raised concerns that his classification was not intended to define stream credits, it was dragooned into service by regulators and mitigation bankers as the core metric enabling the definition of the ES commodity for sale in SMB. In most of the US, a stream credit consisted of a particular number of linear feet of a specific Rosgen stream type as defined in guidance documents for each state jointly prepared by local Corps districts, EPA regions, and state environmental agencies.

While there have long been complaints from Rosgen's critics that his classification system is unavoidably subjective (Doyle, Shields, Boyd, Skidmore, & Dominick, 2007; Juracek & Fitzpatrick, 2003; Roper, Buffington, Archer, Moyer, & Ward, 2008), in practice linear feet of Rosgen stream type proved to be a practical metric, allowing rapid and repeatable measurement of both stream impacts and credits, and thus enabling the stream mitigation market to function smoothly on the basis of a clearly defined commodity for production and sale. By market standards, SMB was a success. By ecological standards, however, there was an increasing sense that while the first decade of stream mitigation banks might have had excellent compliance with standards of performance for channel form, they showed no substantial improvements in other drivers of mitigation: restoration of ecological function and water quality (Bernhardt & Palmer, 2011; Doyle & Shields, 2012). In EPA's terminology, it was not clear that mitigation defined via the Rosgen classification system was providing any "lift."

In interviews, regulators expressed serious concerns with the way the initial Rosgen-based formulation of stream credits overlooked channel scale, creating what they feared were perverse incentives to compensate for damage to larger channels (which are more difficult and expensive to work in because of the quantity and force of the water coming through) with restoration work on headwater or even ephemeral streams. A regulator in one of the very few states that did not initially adopt a credit definition based on Rosgen's classification system explained that decision:

> There were some very valid technical reasons not to go that way. It always comes down to watershed scale: 1,000 feet of the Ohio River is completely different from 1,000 feet of ephemeral channel. ...You could have very similar projects, potentially – putting a bridge across – but a lot more impact. (Author interview, August 3, 2011)

Others learned this lesson through unhappy experience. One Corps staff member in the Pacific Northwest commented:

> You don't want to have people stabilizing large rivers, you know the Yellowstone or Missouri Rivers for example, doing all their work [ecological damage] there but all the easy mitigation is on the tributaries. Then you end up with [an] impaired main stem and tributaries that are restored. That's not ecologically preferable, nor is it good for the public. (Author interview, August 11, 2011)

Likewise, a staff member in charge of revising the guidance for stream mitigation in a Southeastern Corps district noted:

> We had some people actually propose (laughs) ... "I'm going to restore this stream!" And you go out on site and they've got pin flags [sticking out of the dirt] ... that look like a snake on the ground and it's obviously an ephemeral area. So we were trying to avoid that ... conflict we constantly got into. (Author interview, August 11, 2011)

Some interview subjects took this critique farther, raising concerns about whether it was appropriate to define credits based on channel form under any circumstances. Critics often refer to Natural Channel Design, Rosgen's approach, as the "field of dreams" or "build it and they will come" approach because its core premise is that if you build the channel form correctly water quality and ecology will follow (Palmer, Ambrose, & LeRoy Poff, 1997; Sudduth, Hassett, Cada, & Bernhardt, 2011; Violin et al., 2011); critics argue that there is no evidence that this in fact the case. As one regulator in the Southeast put it, "The whole premise of 'build it and they will come' – is it accurate? You know, is this making a difference?" (Author interview, August 29, 2011). And another said, "I get [that it is important to know] what's going to work technically and make sure they won't build a project that's going to blow out [because of improper channel form], but how well will it work [ecologically]? Stability is not everything" (Author interview, August 30, 2011).

Every regulatory jurisdiction in which I conducted interviews had either gone through a major revision of their stream mitigation guidelines in the previous three years to address these concerns or was in the process of doing so. But the catalyst for this national round of revisions was not the concerns of regulatory agency staff members. Most had neither the time nor the local political capital to overhaul their guidance documents. Once a system is in place and clearly understood by the regulated community, it can possess considerable inertia. One Corps staffer described the very slow process of transforming a guidance document based on riparian mitigation to one that addresses broader ecological concerns:

> Well, I think all of us that are biologists would agree that if you do in-stream damage you should completely replace it with in-stream work. You know, but that's not a feasible, realistic approach at this point in where we're at with stream mitigation .... If it were an ideal world, we would have gone straight from riparian buffers to, "You've got to put the stream back like it should be," but that would never happen immediately. It would take years. (Author interview, August 29, 2011)

State regulators made similar comments. For example, the author of a set of stream mitigation guidelines that proposed a new method for measuring debits and credits commented:

> [P]eople are still having a hard time wrapping their brains around this whole concept, but if you go out in the field, you do the measurements, you take a look and see what we're asking for it makes perfect sense. It's just that it's a sea change that we are asking for, so people are resistant. (Author interview, August 3, 2011)

Thus instead of regulators, it was mitigation bankers who catalyzed the current round of deep revisions, largely through the efforts of their trade and lobbying group, the National Mitigation Banking Association (NMBA), within the field of power. The NMBA worked closely with Walter B. Jones, a Republican congressman from North Carolina, the state with the largest mitigation banking industry. Jones made multiple efforts to introduce legislation that promoted mitigation banking, and NMBA newsletters mention Jones' and his efforts frequently and positively. As early as 2001, the NMBA newsletter stated that, "The Association strongly supports the Jones Bill. We have worked with Congressman Jones' office and with the Subcommittee staff on this legislation" (NMBA, 2001, p. 1).

In 2004, after multiple thwarted efforts, Jones finally succeeded, slipping a requirement to revisit federal guidelines for mitigation banking into H.R. 1588, the National Defense Authorization for 2004. When EPA and the Corps released the draft Federal Rule in 2006, then NMBA president Craig Denisoff reminded members, "We need to thank our friends and allies in this effort to improve mitigation standards and improve our industry. First and foremost, Congressman Walter Jones for his tireless efforts to pass legislation to improve banking standards ..." (Denisoff, 2006). When the final Rule was released in 2008, NMBA president Don Ewoldt thanked Congressman Jones for, "being such a staunch ally and friend to our industry" (Ewoldt, 2008), and Ecosystem Marketplace, a web-based clearinghouse for information on market-based environmental management, noted:

> Wetland mitigation bankers hailed their congressional hero, North Carolina Congressman Walter B. Jones Jr., with a standing ovation in Washington DC's marble-

> columned Rayburn House congressional office building just before the April 10th official release of a federal wetland mitigation rule that will revolutionize their business. Through political horse trading and dogged persistence, this Republican congressman, who represents the state with the largest wetland mitigation industry, successfully spearheaded the new rule that spells out a clear preference for wetland mitigation banking while pushing back the industry's competitors. (Kenney, 2008b, p. 1)

The reasons for a standing ovation are clear: the 2008 Federal Rule was a coup for the mitigation banking industry. In her initial report on the Rule for Ecosystem Marketplace, Kenny wrote:

> The new standards encourage the expansion of mitigation banking to provide compensatory mitigation for unavoidable impacts to the nation's wetlands and streams. They create a hierarchy that clearly spells out the agency's preference for this form of mitigation over its current competitors, said EPA Assistant Administrator for Water, Benjamin H Grumbles. ... By emphasizing its preference for mitigation banking, the new rule will likely create a bonanza of business for wetland mitigation bankers, who currently provide only one-third of the mitigation required, according to recent studies. (Kenney, 2008a)

In a second article, published a month later, Kenney quotes then NMBA president Rich Mogensen, who claimed:

> [The Rule] will promote a tripling in the size of the US wetland mitigation banking sector – from roughly 500 banks today to 1500 within the next three-to-five years, with continued growth thereafter. And, with Wall Street gun shy from the nose dive of sub-prime mortgages and hedge funds, deep-pocketed investors are expected to move significant capital into this federally-endorsed green venture. (Kenney, 2008b, p. 1)

The CWA created the market for mitigation banking; mitigation bankers, for their part, worked through the field of power to catalyze substantive changes within the regulatory field in how that legislation has been interpreted and applied. While this is a task of politics and negotiation, it is *also a task of science*, as implementing the new Federal Rule required the development of a new set of commodity-defining metrics.

DEVELOPING METRICS OF ECOSYSTEM FUNCTION WITHIN A CHANGING SCIENTIFIC FIELD, 2008 –

The 2008 Federal Rule, developed within the regulatory field, made a number of important shifts in the interpretation and implementation of Section 404 of the CWA. As mandated in Congressman Jones' section of H.R. 1588, the Rule did indeed raise the status of mitigation banking. Thus far, this seems to be a familiar story of the influence of industry lobbying in regulatory practice. Yet the EPA and the Corps took the opportunity of a mandated overhaul of Section 404 guidance to address many of their outstanding ecological concerns with how mitigation had been implemented to date, valorizing environmental over market-based capital within the regulatory field, and extending a striking do-si-do of agency and market influence that began with the initial passage of the CWA.

Most notably, the Rule shifted away from defining stream credits through proxy measures of linear feet of Rosgen channel type. Instead, the text of the Rule and the response to comments that precedes it in the Federal Register heavily emphasize defining credits and performance standards based on *function*: "the physical, chemical, and biological processes that occur in ecosystems" (Corps & EPA, 2008, p. 19671). Up to that point, the majority of stream mitigation guidance documents measured the ecological lift of mitigation projects by proxy (via relatively easy-to-assess physical characteristics drawn from the Rosgen channel classification). The new Rule, by contrast, pushed strongly for direct measurement of ecosystem processes. Functional assessments were not required because the EPA and Corps acknowledged that in most areas of the country no such metrics were available, but the document was clear that the most common way of defining and measuring mitigation credits was no longer acceptable:

> Some commenters noted that the proposed rule emphasizes functional standards instead of area-based performance standards, and said that it will be difficult for the Corps to move to a functional approach because simple functional assessment methods do not exist for many types of wetlands [or other aquatic resources], and regulators are much more comfortable with measuring acres and linear feet .... Functional standards are necessary to demonstrate that compensatory mitigation projects offset losses of aquatic resource functions resulting from activities authorized by DA permits. Area-based performance standards tied to functions can also be used to determine the functional capacity of a compensatory mitigation project. *However, area or linear measures alone would not constitute ecological performance standards.* (Corps & EPA, 2008, p. 19644, emphasis added)

Unsurprisingly, this shift generated pushback when it was first proposed in 2006. In the response to comments that precedes the text of the Federal Rule, EPA and the Corps noted, "While some commenters supported the use of functional assessments, others recommended retaining replacement ratios based on area until there is an approved model for accurate functional assessment" (*ibid.*, p. 19634). In response to these concerns, EPA and the Corps did offer a practicality rider:

> Functional assessments will be used to determine compensatory mitigation amounts in cases where such methods are available, appropriate, and practicable for use. There are

> on-going efforts to develop and refine functional assessment methods and other science-based assessment tools. If appropriate functional assessment methods are not available, or if it is not practicable to use the appropriate and available functional assessment method for a particular project, then other appropriate metrics are to be used. (*ibid.*, p. 19634)

But every one of the agency staff whom I interviewed was clear that the new Rule was intended to move stream mitigation across the US toward functional definitions of credits, and they were all taking substantive steps in that direction.

The issue was how to create the "science-based" metrics for functional assessment for which the Rule called. The new function-based metrics had to capture something far more complex than simple physical indicators – actual physical, biological, and chemical processes – but in a form that enabled the mitigation market, and thus the CWA, to continue to function. Phrased differently, the burning question was how to satisfy the demands of science, of the regulated community (developers, bankers, etc.), and of policy implementation simultaneously. As one EPA staffer said:

> [T]his is a regulatory program. We cannot be pie in the sky. The most elegant and difficult thing of working at the science-policy interface in a program that gets the scrutiny of this program ... is that there are a lot of constraints from the simple fact that we're running a program: permits have to go out, decisions need to be made .... [I]t can't wait for some nebulous future [scientific breakthrough]. (Author interview, August 20, 2011)

Balancing these demands was further complicated by the unsettled, dynamic state of stream science. As the same EPA staffer quoted above put it:

> I think that has been the difficulty of environmental regulation: because ecology is [a] new, relatively new, science. Ecology was a relatively new science when the Clean Water Act was put in place. Stream restoration is, it evolved in the last two years, let alone the last, you know, 40. And so I think ... there are tensions between ... the temporal scale of politics in this country and how natural systems work and what it means to do effective research to inform something. (op. cit.)

A Corps staff member noted the same difficulty resolving an evolving science with regulatory mandates when he observed that stream restoration is "an evolving science":

> [W]e're trying to do what is currently considered the best available science out there and I think we do a pretty good job of that [but] ... we don't have time to stop our regulatory permit program and say, "Ok everybody, we're just going to stop issuing permits nationwide until we figure out what's the best possible thing we can do when it comes to stream restoration." It just does not work that way. And so, when we're dealing with an evolving science – which this is in my mind – you know, you try to incorporate lessons learned. (Author interview, September 16, 2011)

Given these issues, whoever was responsible for developing the new functional assessment metrics clearly had a difficult task in front of them. The question was who that would be. Strikingly, the regulatory agency staff writing and revising stream mitigation guidance documents in response to the Rule have consistently turned to each other and even more so to paid consultants.

As I and many others have argued, the increasing valorization of privately produced, commercially applicable knowledge is a central characteristic of neoliberal science policies (e.g. Canaan & Shumar, 2008; Lave, 2012c; Lave, Mirowski, & Randalls, 2010; Mirowski, 2011; Nedeva & Boden, 2006). The rise of Rosgen and his Natural Channel Design approach was symptomatic of that shift (Lave, 2012a, 2012b), and the pattern continued with SMB (Lave et al., 2010). Federal and state agencies did not ask academics to develop the new functional assessment tools. While two of the thirteen jurisdictions in which I conducted interviews mentioned direct collaboration with academics, those two are notable in being the only districts, regions, or states I know of that never at any point used Rosgen's approach as a proxy measure for ecological lift. Instead, the scientific and technical expertise to develop new, more complex measures of function was drawn from within federal and state agencies, and even more so from their paid consultants.

In interviews, agency staff often described themselves by their scientific training: "I'm a botanist," "I'm a fisheries guy," or "I am a turtle biologist." The same was true of how people described their colleagues. As one Corps staffer said, "[O]bviously, it's not the [Army] Corps of Biologists, but most of the people in Regulatory are biologists" (Author interview, September 16, 2011). Given this, it is not surprising that interview subjects consistently referred to themselves and colleagues at their own and other agencies as a primary source of scientific expertise in the current major overhaul of regulatory documents on stream mitigation. An EPA staff member described inviting staff from a number of federal resource and regulatory agencies to a workshop in order, "to get the state of the science ..." (Author interview, August 30, 2011). A Corps staffer, when asked if he would change anything about his recently revised stream mitigation guidance document said no because of, "the team that we brought to the table .... I feel extremely comfortable with the technical competencies that they offered up" (Author interview, August 19, 2011). A final example comes from an EPA staff member who reported relying on his colleagues in order, "to be sure the expertise and references being used are the best" (Author interview, August 30, 2011).

In addition to drawing on their own and colleagues' expertise, almost three-fourths of Corps districts, EPA regions, and states reported working actively with consultants to develop the new functional assessment metrics. In all but one case, that consultant was Will Harman, a North Carolina restoration expert, short course instructor, and one of pioneering restoration practitioners in the Southeast. He founded Buck Engineering, a Natural Channel Design-based stream restoration firm, and ran it until shortly after it was acquired by Michael Baker Engineering in 2006. Harman has also taught a set of popular short courses on design, construction, and permitting of Natural Channel Design projects, parlaying his experience into a very visible career in both design and teaching.

When EPA wanted to develop a set of functional assessment metrics to enable implementation of the 2008 Federal Rule, it turned to Harman and his frequent co-instructor, former US Fish & Wildlife Service staffer Rich Starr. In consultation with EPA staff, Harman and Starr and a small group of consultants compiled *A Function-Based Framework for Stream Assessment and Restoration Projects* (Harman et al., 2012). This 344-page document provides an extensive set of metrics and performance standards that attempt to address actual ecosystem functions per the 2008 Federal Rule. The idea is for states to pick and choose among them, crafting a set of metrics and standards appropriate for their specific political and ecological circumstances. As approximately two-thirds of the compiled metrics and standards build directly or indirectly on Rosgen's Natural Channel Design approach, the new approach seems less of a departure than the language in the 2008 Federal Rule would suggest, but is still a clear move to take stream ecology more seriously.

With the backing of both EPA and the Corps, Harman and Starr have begun promulgating their function-based approach from Alaska to South Carolina, reaching out to state and federal agency staff and restoration consultants at four-to-five day workshops that combine classroom education on the Pyramid, the intellectual framework for the *Function-Based Assessment* document, with field techniques from the Rosgen Level I and II short courses. In effect, agency staff at the federal and state levels are replacing one go-to private sector scientist, Dave Rosgen, with another, choosing a strikingly similar format and circulation of knowledge claims: the charismatic consultant and the short course.

CONCLUSION

The current success of SMB in the US is frequently described as evidence of the inevitability of neoliberal approaches to environmental management and the ascendance of market-based capital forms in the regulatory field. From the outside, it seems that command-and-control approaches were replaced smoothly and easily within the regulatory field by a new and thriving market in stream credits, allowing developers to avoid the uncertain and time-consuming process of compliance with the CWA. Yet as I have demonstrated here, the seemingly teleological process of neoliberalizing stream regulation hides a messy and non-linear process whereby agency staff and market actors within the regulatory field have struggled over a fluctuating balance of profit and environmental protection. Far from the triumphant narrative of sweeping regulatory reform, entrepreneurial mitigation banking was jerry-rigged into the rough approximation of a market-based form that nonetheless requires the state not only to organize but also to maintain it.

I have argued that this messy, partial neoliberalization is the product of two key sets of relations: *external* relations between the regulatory field and initiatives in the fields of power and science (specifically, Executive Branch regulatory reform efforts, the ecosystem services paradigm, and the neoliberal valorization of private sector science) that catalyzed the development of entrepreneurial mitigation banking, and relations *internal* to the regulatory field (between agency staff and mitigation bankers) that have shaped mitigation's implementation. Demonstrating the central importance of interfield relations, regulatory reform advocates in the Executive Branch latched onto the ideological and practical tools provided by the ecosystem services framework and appropriated the existing regulatory form of mitigation banking as a medium to force the creation of the command-and-control/neoliberal hybrid that is mitigation banking today. In the latter set of intra-field relations, regulators engaged in a fascinating do-si-do with mitigation bankers, struggling to define the overall direction of SMB, and of environmental regulation more broadly. Taken together, these paired movements underline the importance of extending Bourdieu's fundamentally relational analytical framework to the relations between fields, not just those within them.

NOTES

1. Morgan Robertson hunted down many of the sources in this section.
2. Hess (2011) lays out a similar time line of regulatory reform in relation to electricity deregulation.

ACKNOWLEDGMENTS

This paper benefited greatly from discussions at the Political Sociology of Science Workshop on the Neoliberal State, Science, and Technology organized by Daniel Lee Kleinman, Scott Frickel, David J. Hess, and Kelly Moore, which was held in June 2012 at the University of Wisconsin-Madison. Kelly Moore provided very useful commentary on the first version of the manuscript, as did David J. Hess on the revised manuscript. This paper has also benefited from discussion with audiences at the Geography Department at University College London; the RESTORE workshop in Uppsala, Sweden; the Institute for Environmental Sciences & Technology (ICTA) at the Autonomous University of Barcelona; the workshop on the Political Economy of Research & Innovation at the Sociology Department at Lancaster University, UK; the Dimensions of Political Ecology conference at the University of Kentucky; and the Association of American Geographers. Last and most importantly, this paper is a product of an NSF-funded collaborative research project with Martin Doyle and Morgan Robertson (The Emerging Commodity of Restored Streams: Science, Policy, and Economics in New Markets for Ecosystem Service Commodities (NSF-BCS-0961551)). The opportunity to work with the two of them over the last five years has been a tremendous gift.

REFERENCES

Bernhardt, E., & Palmer, M. A. (2011). River restoration: The fuzzy logic of repairing reaches to reverse catchment scale degradation. *Ecological Applications*, *21*(6), 1926–1931.

Bourdieu, P. (1975). The specificity of the scientific field and the social conditions of the progress of reason. *Social Science Information*, *14*(6), 19–47.

Bourdieu, P. (1983). The field of cultural production, or: The economic world reversed. *Poetics*, *12*, 311–356.

Bourdieu, P. (1991). Genesis and structure of the religious field. In C. Calhoun (Ed.), *Comparative social research: Religious institutions* (pp. 1–44). Greenwich, CT: JAI Press.

Bourdieu, P. (1996a). *The state nobility: Elite schools in the field of power*. Stanford, CA: Stanford University Press.

Bourdieu, P. (1996b). *The rules of art: Genesis and structure of the literary field*. Stanford, CA: Stanford University Press.

Bourdieu, P. (1998). Rethinking the state: Genesis and structure of the bureaucratic field. In P. Bourdieu (Ed.), *Practical reason: On the theory of action*. Stanford, CA: Stanford University Press. (Original Edition, Sociological Theory, 1994.)

Bourdieu, P., & Wacquant, L. J. D. (1992). *An invitation to reflexive sociology*. Chicago, IL: University of Chicago Press.

Bourdieu, P., Wacquant, L. J. D., & Farage, S. (1994). Rethinking the state: Genesis and structure of the bureaucratic field. *Sociological Theory*, *12*(1), 1–18.

Camic, C. (2011). Bourdieu's cleft sociology of science. *Minerva*, *49*, 275–293.

Canaan, J. E., & Shumar, W. (2008). Higher education in the era of globalization and neoliberalism. In J. E. Canaan & W. Shumar (Eds.), *Structure and agency in the neoliberal university* (pp. 1–30). London: Routledge.

Carter, J. (1978). Executive order 12044: Improving government regulations.

Carter, J. (1980, June 13). White House Memorandum for the heads of executive departments and agencies.

Clinton, W. (1993). Executive order 12866: Regulatory planning and review. Washington, DC.

Corps (US Army Corps of Engineers), & EPA (US Environmental Protection Agency). (2008). Compensatory mitigation for losses of aquatic resources. *Federal Register*, *73*.

Costanza, R., d'Arge, R., de Groot, R., Farber, S., Grasso, M., Hannon, B., ... van den Belt, M. (1997). The value of the world's ecosystem services and natural capital. *Nature*, *387*, 253–260.

Cronon, W. (1991). *Nature's metropolis: Chicago and the Great West*. New York, NY: W.W. Norton & Company.

Daily, G. C. (1997). *Nature's services: Societal dependence on natural ecosystems*. Washington, DC: Island Press.

Dempsey, J., & Robertson, M. (2012). Ecosystem services: Tensions, impurities, and points of engagement within neoliberalism. *Progress in Human Geography*, *36*(6), 758–779.

Denisoff, C. (2006). President's Corner. National Mitigation Banking Association.

Doyle, M. W., & Shields, F. D. (2012). Compensatory mitigation for streams under the Clean Water Act: Reassessing science and redirecting policy. *Journal of the American Water Resources Association*, *48*(3), 494–509.

Doyle, M. W., Shields, F. D., Boyd, K. F., Skidmore, P. B., & Dominick, D. W. (2007). Channel-forming discharge selection in river restoration design. *Journal of Hydraulic Engineering*, *133*(7), 831–837.

Ehrlich, P. R., & Ehrlich, A. H. (1981). *Extinction: The causes and consequences of the disappearance of species*. New York, NY: Random House.

Ewoldt, D. (2008). President's Corner. National Mitigation Banking Association.

Eyal, G. (2013). Spaces between fields. In P. Gorski (Ed.), *Bourdieu and historical analysis* (pp. 158–182). Durham, NC: Duke University Press.

Glubiak, P. G., Nowka, R. H., & Mitsch, W. J. (1986). Federal and state management of Inland Wetlands: Are states ready to assume control? *Environmental Management*, *10*, 145–156.

Gomez-Baggethun, E., de Groot, R., Lomas, P. L., & Montes, C. (2010). The history of ecosystem services in economic theory and practice: From early notions to markets and payment schemes. *Ecological Economics*, *69*(6), 1209–1218.

Gomez-Baggethun, E., & Ruiz-Perez, M. (2012). Economic valuation and the commodification of ecosystem services. *Ecological Economics*, *35*(5), 613–628.

Greenwalt, L. (1979). Mitigation in our future. In G. A. Swanson (Ed.), *The mitigation symposium* (pp. 78–81). Fort Collins, CO: Rocky Mountain Forest and Range Experiment Station.

Harman, W., Starr, R., Carter, M., Tweedy, K., Clemmons, M., Suggs, K., & Miller, C. (2012). *A function-based framework for stream assessment and restoration projects*. Washington, DC: US Environmental Protection Agency.

Harvey, D. (2005). *A brief history of neoliberalism*. Oxford: Oxford University Press.

Helliwell, D. R. (1969). Valuation of wildlife resources. *Regional studies*, *3*, 41–49.

Hess, D. J. (2011). Electricity transformed: Neoliberalism and local energy in the United States. *Antipode*, *43*(4), 1056–1077.

Hilts, P. J. (1991). At heart of debate on Quayle council: Who controls Federal regulations? *The New York Times*, December 16.

Horak, G. C. (1979). The status and success of fish and wildlife mitigation in western water resource projects, 1946–1979. In G. A. Swanson (Ed.), *The mitigation symposium* (pp. 391–395). Fort Collins, CO: Rocky Mountain Forest and Range Experiment Station.

Hough, P., & Robertson, M. M. (2009). Mitigation under Section 404 of the Clean Water Act: Where it comes from, what it means. *Wetlands Ecology and Management*, *17*, 15–33.

Joeres, E. F., & David, M. (1983). Is a viable implementation of TDPs transferable? In E. F. Joeres & M. David (Eds.), *Buying a better environment: Cost-effective regulation through permit trading*. Land Economics Monograph No. 6. Madison, WI: University of Wisconsin Press.

Juracek, K. E., & Fitzpatrick, F. A. (2003). Limitations and implications of stream classification. *Journal of the American Water Resources Association*, *39*(3), 659–670.

Kenney, A. (2008a). *New EPA wetland guidelines to boost mitigation banks*. Ecosystem Marketplace.

Kenney, A. (2008b). *Environmentalists sound off on EPA wetland regs*. Ecosystem Marketplace.

King, D. M. (1992). Avoiding another taxpayer bailout. *National Wetlands Newsletter*, *14*(1), 11–12.

King, R. T. (1966). Wildlife and man. *New York Conservationist*, *20*(6), 8–11.

Lave, R. (2012a). Bridging political ecology and STS: A field analysis of the Rosgen Wars. *Annals of the Association of American Geographers*, *102*(2), 366–382.

Lave, R. (2012b). *Fields and streams: Stream restoration, neoliberalism, and the future of environmental science*. Athens, GA: University of Georgia Press.

Lave, R. (2012c). Neoliberalism and the production of environmental knowledge. *Environment & Society: Advances in Research*, *3*(1), 19–38.

Lave, R., Doyle, M., & Robertson, M. (2010). Privatizing stream restoration in the US. *Social Studies of Science*, *40*(5), 677–703.

Lave, R., Mirowski, P., & Randalls, S. (2010). Introduction: STS and neoliberal science. *Social Studies of Science*, *40*(5), 659–675.

Mirowski, P. (2011). *Science-mart: Privatizing American science*. Cambridge, MA: Harvard University Press.

National Performance Review. (1993). Improving regulatory systems. REGO2: Encourage more innovative approaches to regulation. Office of the Vice President, Washington, DC.

National Performance Review. (1995). Reinventing environmental regulation.

Nedeva, M., & Boden, R. (2006). Changing science: The advent of neoliberalism. *Prometheus*, *24*(3), 269–281.

NMBA. (2001). NMBA newsletter. National Mitigation Banking Association.

Norgaard, R. B. (2010). Ecosystem services: From eye-opening metaphor to complexity blinder. *Ecological Economics*, *69*(6), 1219–1227.

Odum, E. P., & Odum, H. T. (1972). Natural areas as necessary components of man's total environment. *Transactions of the thirty seventh North American wildlife and natural resources conference* (Vol. 37, pp. 178–189). Washington, DC: Wildlife Management Institute.

OIRA. (1998). The role of economic analysis in regulatory reform. Report to congress on the costs and benefits of Federal regulations. Washington, DC: Office of Management and Budget.

Palmer, M. A., Ambrose, R. F., & LeRoy Poff, N. (1997). Ecological theory and community restoration ecology. *Restoration Ecology*, *5*, 291–300.

Reagan, R. (1981). Executive order 12291: Federal Regulation. Washington, DC.

Redmond, A. (1992). How successful is mitigation? *National Wetlands Newsletter*, *14*(1), 5–6.

Robertson, M. M. (2004). The neoliberalization of ecosystem services: Wetland mitigation banking and problems in environmental governance. *Geoforum*, *35*(3), 361–373.

Robertson, M. M. (2006). The nature that capital can see: Science, state and market in the commodification of ecosystem services. *Environment and Planning D: Society and Space*, *24*(3), 367–387.

Robertson, M. M. (2013a). Jimmy Carter and the archaeology of market environmentalism, Part 1. Retrieved from http://wetlandia.blogspot.com/2013/07/coase-carter-and-archaeology-of-market.html. Accessed on August 17, 2013.

Robertson, M. M. (2013b). Jimmy Carter and the archaeology of market environmentalism, Part 2. Retrieved from http://wetlandia.blogspot.com/2013/08/jimmy-carter-and-archaeology-of-market.html. Accessed on August 17, 2013.

Roper, B. B., Buffington, J. M., Archer, E., Moyer, C., & Ward, M. (2008). The role of observer variation in determining Rosgen stream types in Northeastern Oregon mountain streams. *Journal of the American Water Resources Association*, *44*(2), 417–427.

Scott, J. C. (1985). *Weapons of the weak: Everyday forms of peasant resistance*. New Haven, CT: Yale University Press.

Short, C. (1988). *Mitigation banking*. Washington, DC: US Fish & Wildlife Service.

Sudduth, E., Hassett, B. A., Cada, P., & Bernhardt, E. (2011). Testing the field of dreams hypothesis: Functional responses to urbanization and restoration in stream ecosystems. *Ecological Applications*, *21*(6), 1972–1988.

US Congress. (1958). Fish and Wildlife Coordination Act (16 USC 661-666). Government Printing Office, Washington, DC.

US Regulatory Council. (1980). Innovative techniques in theory and practice. *Proceedings of a regulatory council conference*, Washington, DC.

Violin, C. R., Cada, P., Sudduth, E., Hassett, B. A., Penrose, D. L., & Bernhardt, E. (2011). Effects of urbanization and urban stream restoration on the physical and biological structure of stream ecosystems. *Ecological Applications*, *21*(6), 1932–1949.

Weidenbaum, M. (1997). Regulatory process reform from Ford to Clinton. In *Regulation*. Washington DC: Cato Institute.

BEEKEEPERS' COLLECTIVE RESISTANCE AND THE POLITICS OF PESTICIDE REGULATION IN FRANCE AND THE UNITED STATES

Sainath Suryanarayanan and Daniel Lee Kleinman

ABSTRACT

This paper utilizes controversies over the role of a set of insecticides in mass honey bee die-offs in two different national contexts – France and the United States – in order to understand the science-state nexus in a comparative manner. On the one hand, the French government in 1999 and 2004 suspended the commercial use of the insecticidal products that beekeepers suspected of causing the honey bee declines. On the other hand, the US government has to date refused to heed beekeepers' calls to limit the usage of the very same set of insecticides. We examine why the governments of France and the United States came to contrasting conclusions regarding broadly similar technoscientific issues. The divergent outcomes, we argue, are not simply the result of predetermined differences in the two states' regulatory paradigms (with France being "precautionary," and the United States adhering to a "sound science" approach), but are underpinned by divergent forms of beekeepers'

Fields of Knowledge: Science, Politics and Publics in the Neoliberal Age
Political Power and Social Theory, Volume 27, 89–122
Copyright © 2014 by Emerald Group Publishing Limited
All rights of reproduction in any form reserved
ISSN: 0198-8719/doi:10.1108/S0198-871920140000027011

resistance. The paper further sheds light on non-state actors' use of science and state to contest state (in)action by analyzing how historically influenced differences in state structures, the relational dynamics of beekeepers' and farmers' organizations, and the epistemic cultures of honey bee knowledge production, shaped different forms of resistance and influence in France and the United States.

Large-scale declines in the populations of crucial insect pollinators, exemplified by phenomena of vanishing honey bees, threaten the agricultural, economic, and environmental well-being of several nation-states (National Research Council, 2007). Most scientists agree that ongoing honey bee die-offs are caused by a combination of factors including particular pathogens, parasites, pesticides, and malnutrition (USDA, 2010). However, knowledge regarding which of these factors play a more prominent role is both uncertain and controversial (Suryanarayanan & Kleinman, 2011). In particular, the role of specific agricultural pesticides has been a major flashpoint between and within groups of researchers, government officials, beekeepers, agro-industry actors and environmental activists. The ways in which this controversy, involving similar sets of issues and actors, has unfolded within different *national* contexts offers a valuable opportunity to understand the nexus between scientific knowledge production and state regulation in a comparative manner. *Here we explore why the governments of France and the United States came to contrasting conclusions regarding broadly similar technoscientific issues.*

In what follows, we first place our work within the context of comparative historical scholarship in science and technology studies (STS) and then provide a schematic comparative history of the French and US cases, telegraphing the central differences across the countries, differences that we will detail more fully in subsequent sections. Next, we trace the collective actions of beekeepers in the controversy over newer kinds of insecticides in our two countries. We follow this with a comparative discussion of the factors that have shaped their respective actions.

COMPARATIVE ANALYSIS IN STS

The history of STS from the 1970s until recent years is dominated by case study analyses often undertaken at a micro-level. Although these studies produced important methodological innovations and theoretical insights into how science is made, they also tended to downplay the role of broader institutional factors in shaping laboratory-level knowledge (Kleinman, 2003) and the contours of scientific controversies (Suryanarayanan & Kleinman, 2013). Our work and that of the political sociology of science in general has pointed to the explanatory role of organizational and institutional factors (Frickel & Moore, 2008; Kleinman, 2003). One method for clarifying the significance of such factors is cross-national comparative analysis (Jasanoff, 2005; Kleinman, 1995; Kleinman & Kinchy, 2003; Wright, 1994), an orientation which has been scattered through the scholarship, even while a micro-oriented sociology of scientific knowledge and laboratory studies dominated discussion.

There are various ways in which a comparative method has been used in STS. In anthropological ethnographies of science and technology, a comparative perspective is frequently implicit but sometimes explicit, and it often provides the basis for a cultural analysis of what specific scientific theories and practices mean in local contexts (e.g., Traweek, 1992). Comparative analysis has also been important for studies of science-related regulation (e.g., Jasanoff, 2005; Wright, 1994). More recently, important comparative studies drawing on institutional traditions in political and historical sociology have focused on how institutional factors can explain variations in epistemic cultures (e.g., Wright, 1994). Among this work is Libby Schweber's *Disciplining Statistics: Demography and Vital Statistics in France and England, 1830–1885* (2006) and Marion Fourcade's *Economists and Societies: Discipline and Profession in the United States, Britain, and France* (2009). Schweber examines the effects of states and related institutions on the development and use of policy-relevant statistics in France and England in the middle of the 19th century. Schweber is attentive to how the statist and anti-statist traditions in France and England respectively shaped the character of statistics in these two nations. In many ways similar to Schweber, Fourcade is interested in how national-level factors shape the differential character and successes of scholarly fields. Both authors are influenced by sociologists like Skocpol, and Fourcade helpfully draws on an array of work in organizational analysis to understand how distinctive institutional logics explain differences in the discipline and profession of economics in the United States, Britain, and France.

Much of this emerging macro-analytic comparative analysis in the sociology of science and technology focuses on the role of states and other organizational structures in explaining policy outcomes or scientific field development. Although there is a body of work in social movement studies that examines comparative political opportunity structures (e.g., on the anti-nuclear movement see Kitschelt, 1986), there is less comparative historical work that focuses on the intersection of social movements, science, and policy (for an exception, see Kinchy, 2012; see also Moore, 2008). Likewise, there is no prominent comparative historical scholarship that seeks to understand how, where, and why research is undertaken in laboratory and field experimentation, and what knowledge counts as legitimate knowledge in shaping policy. In what follows, we build on the tradition of those comparative historical scholars in STS who find inspiration and analytical leverage from scholarship in political, organizational, and historical sociology. We consider how state structures and political culture affected the emergence and efficacy of social movements in the controversies in the United States and France over links between specific insecticides and large-scale honey bee deaths, and we attend to how cultural and organizational differences affected the kind of scientific knowledge viewed as credible and as the appropriate basis for policy decisions.

A DIFFERENT REGULATORY BUZZ IN FRANCE AND THE UNITED STATES

In contrast to the United States, where crop pollination is a major source of income for many professional beekeepers (Burgett, Daberkow, Rucker, & Thurman, 2010) – who rely primarily on managing bees for their livelihoods – in Europe, including in France, most professional beekeepers derive a majority of their income from selling honey and beehive products (Chauzat et al., 2013). Following the summer of 1994, professional French beekeepers, who annually brought their beehives into the blooming sunflower fields of central France to make sunflower honey, began noticing new and alarming behavior patterns in their honey bees, followed by "brutal" beehive losses (La Coordination des Apiculteurs de France, 2000a). "Mad bee disease," as the phenomenon came to be called, caused between 30% and 50% beehive losses compared to the normal 5–10% and was accompanied by a severe loss in honey production. Beekeepers noted that bees foraging on these fields spent "too long" "resting" or persistently "cleaning/scratching" their bodies. They seemed to be "hit by a sort of paralysis," with bees frequently failing to land or take-off in front of their beehive entrances. Based on their field observations, French beekeepers associated the emergence of these symptoms of bee behavior and the related "melting away" of bees in beehives with "sublethal intoxication" by "minute" levels of a new "systemic"[1] insecticide imidacloprid – manufactured by the multinational chemical corporation Bayer – which had recently been introduced in sunflowers as a seed-coating under the brand name of Gaucho®. Beekeepers demanded the suspension of Gaucho® until relevant scientific studies showed that they were safe for honey bees (La Coordination des Apiculteurs de France, 2000a). But results from toxicological studies were mixed. Field and laboratory studies by agrochemical industry scientists indicated that adverse effects from insecticidal residues were seen only at levels several fold higher than those found in the pollen and nectar of treated crops. Bayer's studies suggested that the levels of exposure in crop settings did not pose a risk to honey bees (Schmuck, 1999; reviewed in Maus, Curé, & Schmuck, 2003). By contrast, several laboratory and "semi-field" studies by scientists at the French national institutes of agricultural research found minute quantities of the insecticidal compounds, comparable to those found in pollen and nectar, to have chronic "sublethal" and lethal effects on exposed honey bees (Colin et al., 2004; Decourtye et al., 2001; Suchail, Guez, & Belzunces, 2001). Methodological variability and conflicting experimental results in several of these studies contributed to the production of scientific uncertainty and a lack of consensus on the issue. In 1998, a panel of scientific experts called the Commission de Toxiques (ComTox) belonging to the Ministère de l'Agriculture (Ministry of Agriculture) refused to heed beekeepers' calls for a suspension of the insecticides at issue, given the lack of *conclusive* scientific evidence (Maxim & van der Sluijs, 2007). The ComTox's position was predicated on a preference for definitive results erring in favor of false negatives – an academic institutional norm in fields of experimental biology, wherein a substance that may be harmful is deemed not to be so in the absence of definitive causal evidence. Following a protracted period of beekeeper-led mobilizations and street demonstrations in Paris and in other places in late 1998, the Minister of Agriculture overrode the ComTox's recommendations and invoked the precautionary principle, placing the burden of proof of harmlessness on Bayer and suspending the usage of Gaucho® on sunflowers in early 1999 for two years. In other words, beekeeper-led protests shook the prevalent institutional preference for false negative results, and prompted French policymakers to consider the same scientific evidence from a false positive error orientation, wherein a substance that may not be harmful is considered to be harmful in the light of suggestive evidence. Amidst continuing incidences of "mad bee disease" and renewed street demonstrations by beekeepers, in 2001 the Minister of Agriculture renewed the suspension of Gaucho®. Furthermore, in 2004 the ministry expanded the scope of the suspensions to exclude the use of BASF's Régent TS® (with fipronil as the main insecticidal compound) on sunflowers and Gaucho® on maize for a further three years (Maxim & van der Sluijs, 2007).

In the United States, over a decade later, between 2004 and 2005, beekeepers began noticing a strange new phenomenon, which came to be called "Colony Collapse Disorder" or CCD. CCD-affected bees suddenly abandoned their beehives, leaving them to collapse despite having substantial stores of brood, nectar, and honey. Beehive losses increased to between 30% and 90% in several professional beekeeping operations, especially those involved in the business of crop pollination, compared to the prior average of 15% (USDA, 2007; van Engelsdorp, Hayes, & Pettis, 2008). Several long-time beekeepers noted a consistent correlation between CCD and the proximity of their beehives to crops treated with newer kinds of systemic insecticides, such as imidacloprid (e.g., Gaucho®) and clothianidin (e. g., Poncho®). Based on their systematic, if something less than formal, field observations, commercial beekeepers posited that chronic, dietary exposure of their bees to low levels of these chemicals was leading over time to CCD. Accordingly, beekeepers called upon the federal regulatory authority, the US Environmental Protection Agency (EPA), to suspend the commercial usage of the newer systemic chemicals until chronic toxicity studies showed them to be safe to honey bees (National Honey Bee Advisory Board, 2009a). However, several academic and industry scientists pointed to the lack of clear causal evidence from laboratory and field (toxicological) studies (e.g., Ratnieks & Carreck, 2010), and the EPA has to date considered the absence of definitive causal evidence as a scientifically sound justification for not discontinuing the commercialization of the insecticides in question. The EPA's position regarding the newer systemic insecticides, similar to the original position taken by the French ComTox, is based on an institutional preference for a false negative error norm (Suryanarayanan & Kleinman, 2011).

Bayer, "scientific" beekeepers, EPA regulators, and some university scientists maintain that the French outcome was the result of "politics," a reflection of populist pressure in the absence of any clear scientific evidence from field studies; by contrast, according to these stakeholders, the US decision was an outcome of incremental scientific knowledge from field experiments that showed no strong support for the alleged links (e.g., EPA, 2008b; Maus et al., 2003; Oliver, 2011). An alternative explanation would assert that the difference in outcomes was the result of divergent overarching paradigms of risk assessment, themselves based on differing political and cultural environments (Frickel et al., 2010). In this reading, France exercised the precautionary principle in keeping with the European Union's guide to environmental policy in the Maastricht Treaty of 1992. The US decision, on the other hand, was the result of a sound science paradigm, an approach that is based on the assumption that value-based politics and fact-based science are neatly separable (Oreskes & Conway, 2010). Although there are recognized differences in regulatory culture, this explanation obscures the social conditions that facilitate the (potentially opportunistic) invocation by powerful actors of one or the other policy paradigms and the related error norms that underlie them. In France, the Ministry of Agriculture's precautionary re-orientation regarding Gaucho® was made in the face of non-precautionary risk assessments by experts for whom there was no conclusive scientific basis on which to suspend usage of the chemicals. Far from being a standard policy decision, it is regarded as the first environmental issue where France invoked the precautionary principle (Maxim & van der Sluijs, 2007). Similarly, in the United States a substantial historical precedence for precautionary actions regarding chemical regulation meant that the issue of insecticide regulation in the CCD controversy did not automatically fall under a "sound science" paradigm (Brickman, Jasanoff, & Ilgen, 1985; Jasanoff, 1986, 1990).

Thus, neither agrochemical industry actors' and scientists' studies nor regulatory orientations (which broadly favor agribusiness interests in both places) adequately explain the difference in outcomes, because they were similar across both cases. Instead, we argue that the divergent policy outcomes in France and the United States are better explained by differences in *beekeepers' collective resistance* in combination with differences in epistemic cultures of honey bee biology, relations between beekeepers' and farmers' organizations, and historically influenced differences in policymaking structures. In both places, organized groups of beekeepers (re)interpreted scientists' findings in the light of their own knowledge of bees, and they disseminated substantive critiques against agro-industry and the state's scientized policy justifications for commercializing the insecticides. In France, the beekeepers' collective resistance became a unified and sustained social movement replete with media campaigns, strikes, and street protests, whereas in the United States it constituted a motley set of relatively isolated and limited collective actions. Beekeepers' strategic actions and influence in each place were in turn shaped by various factors: contrasting cultures of honey bee research, which led scientists to adopt different evidentiary norms and emphasize different aspects of the available scientific evidence; relational dynamics within and between groups of relevant stakeholders (beekeepers and farmers); and historically influenced structures of policy-making, which institutionalize particular forms of interaction between state and non-state actors. In this context, French beekeepers were able to prompt policymakers and scientists to alter the way in which they handled the knowledge about bee deaths that all parties accepted, taking the uncertainty about ultimate causes as reason to remove Gaucho® and Régent TS® from the market, while the US beekeepers have been unable to reverse the governmental preference for non-precautionary risk assessments that err toward false negative results in the face of uncertainty.

FORMS OF BEEKEEPERS' ORGANIZED RESISTANCE AGAINST STATE ACTORS

Mad Bee Disease in France

Following French beekeepers' demands for the immediate investigation of the links between Gaucho® and "mad bee disease," Bayer's researchers conducted a series of field and "semi-field" tests between 1995 and 1997 (Maxim & van der Sluijs, 2007). On the basis of these studies, Bayer concluded that the permitted levels of Gaucho® used in sunflower seeds were harmless to bees (Schmuck, 1999; reviewed in Maus et al., 2003).

But the incidence of mad bee disease continued unabated in French sunflower fields and expanded to other areas in France. In response, affected beekeepers lobbied the French Ministry of Agriculture for support and for the suspension of Gaucho® on precautionary grounds. By late 1997, the Ministry of Agriculture's Commision de Toxique (ComTox) – a body composed predominantly of experts in (eco)toxicology that is in charge of analyzing the authorization of the entire range of pesticide products (Maxim & van der Sluijs, 2007) – asserted that based on the available evidence (Bayer's research, beekeepers' field analyses) it was not possible to conclude whether or not Gaucho® was responsible for causing mad bee disease (Belzunces & Tasei, 1997).

Beekeepers demanded further research, and so in 1998 the French government initiated an extensive series of laboratory and field studies into the potential role played Gaucho®. These were also sponsored by the European Union, and 5% of the funds came from Bayer (MacIntyre, 1998). In contrast with Bayer's studies up to this point, which suggested that imidacloprid was not present in the pollen and nectar of sunflowers

(Maxim & van der Sluijs, 2007) participating scientists at the national research institutes INRA and CNRS (Institut National de la Recherche Agronomique, or French National Institute for Agricultural Research, and Centre National de la Recherche Scientifique, or National Centre for Scientific Research) found that the pollen and nectar of treated sunflower plants had minute levels of Gaucho® (Bonmatin et al., 2003), which caused sublethal effects on exposed bees in laboratory and "semi-field" studies (e.g., Colin et al., 2004; Decourtye et al., 2001; Suchail et al., 2001).[2] These results were not, however, the final word. They were not corroborated under field conditions, and consequently in late 1998 the ComTox stood by its initial determination, concluding that there was no definitive evidence to support a causal role for Gaucho®. The ComTox called for no regulatory change, but recommended further research (Maxim & van der Sluijs, 2007).

Meanwhile, the three national-level beekeepers' associations (the UNAF, Union Nationale de l'Apiculture Française; SPMF, Syndicat des Producteurs de Miel de France; and SNP, the Syndicat National d'Apiculture) were following these developments closely. Organized under the banner of "The Coordination of Beekeepers in France," they publicly disseminated their own reports and press releases analyzing the scientific studies by Bayer, the Ministry of Agriculture, and the state's research institutes (e.g., La Coordination des Apiculteurs de France, 2000a, 2000b, 2002). Following the ComTox's (in)decision in 1998, the national beekeepers' associations staged a series of nation-wide street protests in December 1998, in alliance with environmental groups and certain farmers' groups, calling for the removal of Gaucho® (Coordination des Apiculteurs dc France, 1998). These eye-catching street demonstrations, including nearly 2,000 beekeepers marching in Paris' streets and gathering by the Eiffel Tower, elicited an almost immediate response from the state. The French beekeepers' stance was publicly supported by farmer organizations such as the Confèdèration Paysanne, environmental activist groups, some honey bee scientists, the Green political party (Les Vertes), and by the then Minister of Environment (Maxim & van der Sluijs, 2007). In the face of beekeeper mobilization, on January 22, 1999, French Minister of Agriculture Jean Glavany overruled the ComTox and suspended the use of Gaucho® on sunflowers for a period of two years, until further research could bring more clarity. The French beekeepers' protests had shaken the taken-for-granted position accepted by governmental officials about uncertainty and error norms. Uncertain knowledge became knowledge enough to act to protect honey bees.

Bayer CropScience immediately protested the government's precautionary decision, and along with other transnational agroindustrial firms such as BASF and Monsanto, sued the Ministry of Agriculture at the highest judicial level, the State Council (Maxim & van der Sluijs, 2007). Here, Bayer argued for the primacy of field studies over lab studies and noted that the former showed no clear evidence of Gaucho®'s role. The national beekeepers' associations entered the lawsuit in support of the Ministry of Agriculture and argued that laboratory studies were equally relevant. In December 1999, the State Council ruled in favor of the ministry and the beekeepers and rejected Bayer's bid to cancel the suspension (La Coordination des Apiculteurs de France, 2000a).

Despite the prohibition of the use of Gaucho® on sunflowers, reports of huge beehive losses in sunflowers continued beyond 1999. In press releases, trade journal articles and protest rallies, beekeepers posited three explanatory scenarios (e.g., La Coordination des Apiculteurs de France, 2000a, 2002):

- Another insecticide Régent TS® (fipronil) produced by the German firm BASF had replaced the suspended Gaucho® on sunflower seeds. Beekeepers observed similar problems in their bees after they foraged in Régent TS®-treated sunflower fields. Additionally, the INRA's scientists were finding fipronil to be at least as toxic to bees and as systemic in plants as imidacloprid (e.g., Colin et al., 2004).
- The continuing incidents of mad bee disease could also be the result of accumulated imidacloprid persisting in soils and plants from previous years of treatment (La Coordination des Apiculteurs de France, 2000b).
- The bees continued to be exposed to Gaucho® through the pollen of treated maize crops, which were flowering in the same areas and at the same time as sunflowers. The matter of maize was especially fraught because it is one of the most economically profitable crops in France.

Analyzing emerging scientific findings, French beekeepers concluded:

> It is evident that during the flowering period of Gaucho® treated sunflower and maize as well as all crops with apical interest contaminated by previous Gaucho® treatments, foraging bees are effectively exposed to doses of imidacloprid which in laboratories have revealed a negative impact. (La Coordination des Apiculteurs de France, 2000b)

By 2000, beekeepers were calling for the permanent removal of imidacloprid *and* fipronil from *all* crops. In October and December of that year, beekeepers and their allies organized another series of well-publicized street demonstrations in several parts of France, along with publications and press releases. On December 18, 2000, for example, beekeepers wearing their characteristic hats, veils, and overalls led crowds through the streets of Paris overlooking the Ministry of Agriculture. The protesters held up beekeeper-figurines and banners with skull and cross bones that stated "Le Morte de L'Apiculture, Retrait Immediat du Gaucho®" (The Death of Beekeeping, Remove Gaucho® Immediately) and "Assassins de la Terre Arretez le Massacre!" (Assassins of Earth Stop the Massacre!).

Soon after the beekeeper-led protests, in February of 2001, the Ministry of Agriculture renewed the suspension of Gaucho® on sunflowers for two more years (Ministère de l'Agriculture, 2001), stating that scientific findings "raised suspicions about the effects of the product, without formally proving its responsibility" (quoted in Maxim & van der Sluijs, 2013). This policy orientation was based on a false positive error norm, which gave serious consideration to evidence that was suggestive but not definitive. However, this precautionary orientation taken by the Ministry of Agriculture was a limited and tentative one that by no means signaled an across-the-board shift in the French government's chemical policymaking. Indeed, the Ministry refused beekeepers' calls to suspend the market authorizations of Gaucho® on maize, based on the ComTox's non-precautionary risk assessments, according to which the treatment of maize seeds with Gaucho® had no link to the mad bee disease (Ministère de l'Agriculture, 2001).

The Coordination of French Beekeepers responded by disseminating its own analysis. The organization concluded that "the suspension of Gaucho® as used as a seed treatment on sunflower seeds has not actually allowed lowland beekeeping to recover" because "Gaucho® treated maize, present in the [same] areas of cultivation also had a flowering period at the same time as that of the sunflowers." That their bees were foraging on flowering maize as well was a crucial fact that they suggested was "generally unknown" (La Coordination des Apiculteurs de France, 2002). French beekeepers thus drew the attention of policymakers and scientists to data that placed the issue of honey bee exposure to Gaucho® in a new light.

The year 2003 was marked by further protests and street demonstrations by beekeepers and their allies. On January 22, 2003, beekeepers marched in Paris against Minister of Agriculture Hervé Gaymard's decision to maintain the authorization of Gaucho® on maize. Frustrated beekeepers were involved in a series of pushes and pulls with the Ministry of Agriculture and the agrochemical industry regarding imidacloprid and fipronil. In April, the farmers' group, Confèdèration Paysanne, called on "all farmers" to "boycott" fipronil and imidacloprid coated seeds (Confèdèration Paysanne, 2003).

Following the beekeepers' protests of 2001, the Ministry of Agriculture had also established an ad hoc Scientific and Technical Committee (CST) comprised mostly of honey bee scientists from various research institutes of the government to look into the "multifactorial" causes of the continuing honey bee declines (Ministère de l'Agriculture, 2001). In assessing the risk posed to honey bees by Gaucho®-treated sunflower and maize, the CST's scientists considered accumulating evidence from laboratory, field, and semi-field studies conducted by agrochemical industry and state scientists. Even though the evidence examined by the CST's scientists was similar to what regulatory authorities and scientists in the United States had at hand, in September 2003, they concluded that Gaucho® in *sunflowers and maize* posed risks of significant concern to honey bees (CST, 2003). The CST scientists' conclusion was shaped by an institutionalized skepticism in the epistemic culture of French honey bee research of field toxicity experiments, which were considered by French bee scientists to be too prone to fluctuations, and hence unreliable (CST, 2003; Maxim & van der Sluijs, 2013). In keeping with an institutional preference for more tightly controlled experiments, the CST's bee scientists constructed risk scenarios that relied more on results from laboratory and "semi-field" studies than from suspect field toxicity experiments (CST, 2003). Vindicating the beekeepers' position, these government scientists' estimated risk scenarios were "in agreement with field observations reported by several beekeepers" in areas of intensive agriculture of maize and sunflower (CST, 2003, p. 11). The beekeepers' associations sought to gain legitimacy and general support for their observations regarding Gaucho® and Régent TS® by eagerly echoing the CST's conclusions (UNAF, 2003), but the Ministry of Agriculture remained aloof. In a tersely worded statement on October 30, 2003, the Minister pointed out that the CST report was "interim" and not "final," and raised further doubts about the validity of the findings in support of beekeepers' observations (Ministère de l'Agriculture, 2003).

With beekeepers continuing to reel under massive honey bee losses perceived to be due to the newer systemic insecticides, affected beekeepers in the Midi Pyrénées region and one of the three national organizations, the UNAF, brought lawsuits against BASF and its Régent TS®. UNAF also sued the Ministry of Agriculture for refusing to suspend Gaucho® on maize (UNAF, 2004a). As these trials played out and received media attention, the public pressure to suspend Régent TS® also increased. In early 2004, beekeepers and their allies renewed street protests. On February 11, 2004, a decade after beekeepers began to see signs of "mad bee disease" in France, the Ministry of Agriculture announced that it had decided to proceed with the formal procedures for withdrawal or suspension of Régent TS® because "the information provided cannot conclude an acceptable risk for many species (birds, mammals, aquatic organisms, *bees*) …" (Ministère de l'Agriculture, 2004; our emphasis). Beekeeper pressure had prompted the government to alter the error and risk norms on which it had previously relied, and on May 25, 2004, the Ministry of Agriculture suspended Gaucho® on maize as well as sunflowers (UNAF, 2004b). This decision notwithstanding, the controversy remained alive after 2004. Although beekeepers' organizations claimed that there was a drop in the unusual incidents of summer mortality that had characterized mad bee disease and a partial and gradual recovery of their beehives, the agrochemical firms Bayer and BASF along with their allies asserted that the governmental suspensions of their insecticides had failed to stymie mad bee disease. Gaucho® continued to be applied on sugar-beets, wheat, and barley, while the use of its active ingredient (imidacloprid) expanded onto pollinator-dependent fruit crops (Maxim & van der Sluijs, 2013). The controversy took a new turn after 2008, when, to the consternation of French beekeeping organizations, the French government granted the agrochemical firm Syngenta authorization to use a related systemic insecticide, thiamethoxam, as a seed-coating (Cruiser®) on maize and rapeseed (Canola). Amidst reports of renewed and widespread bee deaths, French beekeepers sued the Ministry of Agriculture for its decision to authorize Cruiser® (UNAF, 2011) and again took to the streets en masse in October 2011 (Gruzelle, 2011). Following closely on the heels of beekeeper-led agitations, the publication of a study by French bee scientists in the high profile journal *Science* vindicated the beekeepers' causes for concern regarding Cruiser® (Henry et al., 2012), and in June 2012, the French Ministry of Agriculture banned the use of Cruiser® on Canola. More recently, the European Union has voted to impose the world's first ever continent-wide moratorium for two years on the use of imidacloprid, thiamethoxam, and clothianidin, due to their potential links to mass declines of honey bees and other insect pollinators.[3]

In the French case, beekeepers' organized resistance developed into a coherent social movement characterized by sustained and disruptive direct action such as street protests and political demonstrations, legal actions, media and public relations campaigns, and lobbying efforts, all of which were directed against a single state entity – the Ministry of Agriculture. Their movement spurred scientists in agro-industry and the state to initiate a new set of scientific studies into the potential links between mad bee disease and the newer systemic insecticides. Significantly, the unified stance of all three national-level beekeepers' associations and their alliances with groups of farmers, environmentalists and public actors instigated a shift in the responses of government officials and scientists alike to what knowledge and ignorance mattered for making policy regarding Gaucho® and Régent TS®, and this alliance was, as a result, crucial in eliciting *and* sustaining pressure for precautionary policy-level reforms, albeit only for certain products on certain crops.

Collective Action in the Case of CCD in the United States

As we noted, unlike France where commercial beekeeping operations are primarily oriented toward honey production, in the United States a significant portion of the income for commercial beekeepers is generated by the work their bees do in pollinating farmers' crops (e.g., see Burgett et al., 2010). Not surprisingly, then, CCD was first noted by commercial beekeepers in the course of their operations for crop pollination.

As of 2002, the approximately 9% of beekeeping operations that manage close to 90% of all beehives in the United States constitute a heterogeneous community that primarily drives the US honey bee industry (Daberkow, Korb, & Hoff, 2009). They are crucial cogs in the industrial agriculture economy of the United States, which, over the past 60 years or so, has come to place an increasingly unsustainable demand on beekeepers' pollination services (Spivak, 2010). Commercial beekeepers routinely truck several thousand beehives all year around, renting them out to farmers in crop settings all across the United States. In the process, their bees are exposed to a plethora of environmental stressors, including parasitic mites, farmer- and beekeeper-applied chemicals, nutritional constraints, and pathogens.

Several commercial beekeepers with decades of migratory beekeeping experience consistently noted a correlation between the occurrence of CCD and the proximity of their beehives to fields treated with relatively new systemic insecticides such as imidacloprid and clothianidin. Affected beekeepers reported that CCD occurred in colonies several months after initial exposure to the insecticides. It suggested to them that foraging bees, instead of dying immediately (as experienced in bee-kills resulting from exposure to more traditional pesticides), were bringing back to their beehives pollen and nectar contaminated with low levels of the systemic insecticide. This, the beekeepers surmised, had long-term and progressively degenerative effects on developing bees that were chronically exposed to insecticidal stores inside beehives (National Honey Bee Advisory Board, 2009a).

In the face of CCD, with their livelihoods at stake, commercial beekeepers have led several initiatives to counter the use and encourage more stringent regulation of the newer systemic insecticides. In 2003, a group of commercial beekeepers initiated a federal class action lawsuit against Bayer CropScience for incurring unusually high honey bee losses, allegedly due to the proximity of Canola crops, the seeds from which had been treated with Gaucho®, to their beehives. In 2008, after a five-year period involving laboratory tests of pesticide residues in bee samples and testimony from honey bee experts and counter-experts, the judge threw out the beekeepers' case (U.S. District Court Pennsylvania, 2008). Late in that same year, a group of seven veteran commercial beekeepers, belonging to the two national-level beekeepers' organizations (American Beekeeping Federation and American Honey Producers Association), formed the National Honey Bee Advisory Board (NHBAB). These beekeepers, who had methodically studied CCD in their operations and had experienced pesticide-related bee die-offs, sought to make changes in the way the newer systemic insecticides and other (agricultural) pesticides were being regulated and enforced, at both policy- and grass-root levels. Their intent was "to accomplish this through dialogue of interested parties, as well as legal, legislative, and political means" (National Honey Bee Advisory Board, 2009b). In letters, and at (informal and formal) meetings, they voiced their collective concerns to grower groups, beekeeper colleagues, academic and federal honey bee scientists, pesticide manufacturers, and state officials in the US Congress and EPA (see National Honey Bee Advisory Board, 2009a).

The NHBAB's beekeepers pointed out flaws in the federal agency's regulatory assessment of the environmental risks posed by the newer systemic insecticides to honey bees. They critiqued the methodology of toxicological laboratory and field studies and the associated interpretations on the basis of which scientists, regulators, and agrochemical industry actors justified the continued usage of the chemicals. For example, the NHBAB's beekeepers pointed out that the EPA's toxicity tests were premised on an "overreliance" on standards that were developed for "traditional" insecticides with a relatively rapid and lethal "mode of action." Beekeepers were concerned that the EPA's scientists were overlooking plausible sublethal, chronic, and synergistic effects of the newer chemicals on the honey bees' life cycle, a scenario that their own studies on CCD suggested. The NHBAB's beekeepers also noted the fundamental "conflict of interests" arising from having "toxicity studies paid for by companies with vested financial interest in getting the product registered" (National Honey Bee Advisory Board, 2011). In calling for a suspension of imidacloprid-like chemicals, the NHBAB's beekeepers recognized that "the problems facing EPA regarding pesticide regulation will not go away with the banning of one compound" and that the "very regulatory system itself needs reforming" (National Honey Bee Advisory Board, 2009a).

But beekeepers' on-the-ground evidence was dismissed as anecdotal and thus invalid by the EPA, several academic scientists, agro-industry representatives, and some beekeepers. Pointing to the lack of clear, causal scientific evidence from field toxicity experiments on whole beehives, which are considered to be a more definitive means of assessing the "real world" effects of imidacloprid-like insecticides on honey bees in the epistemic culture of US honey bee research (e.g., Johnson et al., 2010), these actors rejected the NHBAB's calls for suspending the insecticides of concern (e.g., EPA, 2008a). In response, the NHBAB stuck to its strategy of dialogue with the EPA and lobbying in Washington. Their exhortations, urging other beekeepers to become active, entailed *individual* actions such as "contacting your representatives in Washington" and "supporting additional funding of independent toxicity research" (National Honey Bee Advisory Board, 2011). Crucially, even though the NHBAB claimed to "represent" the two national beekeepers' associations, neither association lent its official support to the NHBAB's cause. To the contrary, the NHBAB's actions were undercut by an influential minority of dissenting "scientific" beekeepers, and by commercial beekeepers, who blamed beekeeper practices for CCD and viewed the newer systemic insecticides as heaven-sent compared to the more traditional insecticides.

In the face of a divided community of beekeepers, the NHBAB enrolled, on a part-time basis, a leading academic insect toxicologist and honey bee scientist, James Frazier, as their "scientific advisor." In 2009, the NHBAB's beekeepers even agreed to collaborate with Bayer researchers, who sought their assistance in a field study intended to provide the EPA with realistic information about the toxic effects of Bayer's latest systemic insecticide, Movento® (with the active compound spirotetramat), on beehives. However, the so-called cooperative field study ended bitterly, with Bayer researchers suggesting that Movento® had only benign effects on bees and participating beekeepers asserting that the data pointed to more pernicious effects.

The NHBAB's beekeepers also made alliances with environmental groups, such as the National Resources Defense Council, Beyond Pesticides, and Pesticide Action Network, which have historically mobilized against pesticide contamination and related issues. Beekeepers took part in the environmental groups' media campaigns against Bayer CropScience's systemic chemicals and against what they perceived to be willful negligence on part of the EPA. In part, based on their concerns, environmental groups launched a lawsuit against the EPA to overturn the "conditional registration" it had granted in 2009 for Movento®. In this case, a US district judge ruled in favor of the environmental groups on the basis of the EPA's "procedural failure" to provide adequate notice and opportunity for public comment on Bayer's application for the pesticide's registration (U.S. District Court, 2010). However, by late 2010, the EPA had re-instated Movento®'s "conditional registration" status, having seemingly surmounted their administrative failures (EPA, 2010).[4]

In this instance, we see that commercial beekeepers never became mobilized as a coherent and powerful wall of collective resistance. Instead, they remained a hodge-podge of "dialogues," lobbying efforts, public relations, and litigation. They failed to become a coherent movement and never developed the sorts of "contentious challenges" or disruptive politics that are thought to characterize successful social movements (Tarrow, 2011). They were unable to prompt government officials to rethink the false negative error norm on the basis of which they make policy or to reconsider when the absence of knowledge – ignorance or uncertainty – justifies precautionary regulation of agrochemicals.

That said, the increased public pressure brought about by the NHBAB's actions led the EPA to signal its shared concern for pollinator and pesticide issues by forming a "Pollinator Protection Workgroup" in 2011. The workgroup to date seems to have focused its efforts on developing better warning labels and better educating pesticide users about risks to pollinators. Also in 2011, agrochemical industry scientists and EPA officials met to explore data requirements for taking into account sublethal effects and effects on honey bee brood. While this may have come about in part due to the pressure brought by beekeeper and environmental groups to devise better toxicity studies, it was also in the transnational agrochemical corporations' interests to move toward a globally harmonized and streamlined process of pesticide risk assessments (see Fischer & Moriarty, 2011).

Thus, while beekeepers in the United States could not bring about policy-level shifts, they were, to a limited extent, able to affect public perception with the aid of environmental activist groups and to push regulatory and agrochemical industry actors to reckon more seriously with their agenda regarding the newer systemic insecticides. Of course, beekeepers' struggles in both France and the United States had some overlapping elements. They produced substantive critiques of the regulatory governance and scientific studies related to the systemic insecticides, which addressed broader systemic flaws in the regulatory system of pesticide risk assessment as well. In doing so, both French and US beekeepers enrolled scientists, inspired the execution of new studies, and utilized emerging scientific findings to furthering their claims. But there were also important differences in the beekeepers' actions of resistance. Most crucially, French beekeepers mounted a cohesive and sustained social movement, while US beekeepers did not.

In the following section, we discuss the factors that shaped the fields of resistance in which beekeepers struggled to get particular agrochemicals removed from commercial usage in the United States and France. Our argument, first, is that the "epistemic culture" (Knorr-Cetina, 1999) of honey bee science in France placed a greater weight on the suggestive results of laboratory experiments and gave much less credence to the sorts of "real world" field experiments that agrochemical industry scientists and academic scientists elsewhere were pushing for. Second, all three national-level beekeepers' associations in France were unified in their opposition to the newer systemic chemicals, notwithstanding the heterogeneity among their individual members. At the same time, French agro-industry groups were more divided, with several influential farmers' groups organizing alongside beekeepers in rallies against the supporters of the newer systemic chemicals. Third, the centralized structure of French regulatory policymaking and its general lack of access to institutional engagement by civil society groups rendered it more vulnerable to directed extra-institutional efforts of beekeepers' social movements. All of these factors led the French government and allied scientists to develop knowledge regarding Gaucho®'s effects on honey bees from a false positive standpoint and use this knowledge to undergird a precautionary policy (Table 1).

On the other hand, a historically influenced convergence in the United States between the norms of academic honey bee toxicology and the EPA's regulatory orientation has meant that both policymakers and scientists tend to accept false negative results from field experiments as the gold standard for considering any shift in the regulatory policy regarding the newer systemic insecticides (Suryanarayanan & Kleinman, 2013). Moreover, the beekeepers' movement in the United States was fractured from within by dissonance among various groups of influential beekeepers. And the decentralized and permeable structure of regulatory policymaking in the United States further served to diffuse and thus weaken beekeepers' attempts to remove imidacloprid and similar chemicals. In this context, the taken-for-granted error norms accepted by academic scientists, government officials, and chemical companies were never fundamentally challenged, and thus,

Table 1. Key Factors That Shaped the Different Beekeepers' Movements and Policy Outcomes in France and the United States.

National Context	Toxicological Epistemic Culture	Bee-Industry Groups	Agro-Industry Groups	Policymaking Structures	Beekeepers' Resistance	Policy Outcome
France	Preference for laboratory studies	More unified	More divided	Centralized, closed	Targeted, conflict-based	Precautionary (false positive)
United States	Preference for field studies	More divided	More unified	Decentralized, permeable	Dispersed, consensus-based	Non-precautionary (false negative)

in the face of uncertainty about the role of systemic insecticides in contributing to CCD, these chemicals remained on the market (Table 1).

Epistemic Cultures of Honey Bee Biology

An array of laboratory experiments (several by French and European bee scientists) provided suggestive and sometimes conflicting evidence linking exposure from low levels of newer systemic insecticides like imidacloprid to behavioral, motor, and physiological abnormalities in exposed honey bees (reviewed in Blacquière, Smagghe, van Gestel, & Mommaerts, 2012; Desneux, Decourtye, & Delpuech, 2007). However, these results were not supported *conclusively* by results from field experiments conducted by university, government and chemical industry scientists.[5] Bayer scientists argued that the field experiments, and not laboratory experiments, were a more accurate and definitive way to assess the purported existence of links between their insecticides and honey bee deaths (e.g., Maus et al., 2003). The agrochemical industry's emphasis on the decisiveness of field experiments resonated in contrasting ways within the scientific and regulatory communities of France and the United States. While academic and regulatory agency scientists in both places clearly recognized field settings to be crucial for assessing the risks posed by the concerned insecticides to honey bees, they differed in the extent to which they lent weight to field experiments vis-à-vis laboratory experiments.

The French CST's honey bee scientists considered the field experiments conducted by agrochemical industry and government scientists alike to be insufficiently valid for assessing the risks posed by Gaucho® to honey bees in sunflower and maize (CST, 2003). They were much more reliant on the results of laboratory and semi-field studies on artificially caged bees and miniaturized beehives. The CST's evaluation criteria reflected the formal and informal norms of European honey bee research, where scientists have tended to be skeptical of the validity of the results from field experiments due to the huge amount of variability that such experiments grapple with (Maxim & van der Sluijs, 2013; personal communication: Marla Spivak, 2010). Field tests of the effects of a systemic agrochemical on beehives are seen by bee scientists to raise two problems: the interference of other ambient chemicals and environmental factors in the tested chemicals' influence and the challenge of ensuring a representative exposure to the tested chemical.

In contrast to the non-precautionary risk assessment frameworks of the ComTox, which according to Maxim and van der Sluijs (2013) had only one acting honey bee scientist member, and Bayer, the CST developed a new approach which incorporated "*various uncertainties* due to extrapolations from single-species *laboratory data* to a multi-species ecosystem" (Halm, Rortais, Arnold, Tasei, & Rault, 2006, p. 2450, our emphasis). Seen in this light, while the beekeepers' social movement in France was a game changer that prompted policymakers to limit the usage of Gaucho® and convene the CST, they would probably not have been able to sustain and expand the suspension of the insecticides without the CST's imprimatur, which itself was shaped by a culture of institutionalized skepticism regarding the reliability of field toxicity tests on honey bees.

By contrast, in the United States, academic and regulatory cultures concerned with honey bees and pesticides concur on the primacy of field experiments as the more definitive arbiter for resolving any purported link between honey bee deaths and newer systemic insecticides. In a 2010 review of pesticides and honey bee toxicity, academic honey bee scientists wrote the following:

> acute toxicity tests on adult honey bees may be particularly ill-suited for the testing of systemic pesticides because of the different route of exposure bees are likely to experience in field applications. Chronic feeding tests using whole colonies *may provide a better way to quantify the effects of systemics.* (Johnson et al., 2010; our emphasis)

However, the prevalent field experimental approach ignores the cumulative, interactive and sublethal effects of low levels of systemic insecticides in combination with other ambient factors on honey bee health (Kleinman & Suryanarayanan, 2013). Consequently, some bee scientists at US universities and at the USDA are turning to laboratory settings, where it is easier to set up controlled tests for studying multifactorial interactive effects. In a cutting-edge study that combined field and laboratory research, the USDA's chief bee scientist Jeff Pettis, and his collaborators, found that when newly emerged honey bees from field beehives fed with extremely low levels of the systemic insecticide imidacloprid were exposed in the laboratory to a common bee pathogen, they were much more likely to develop infections than were young bees from untreated beehives (Pettis, vanEngelsdorp, Johnson, & Dively, 2012). However, the EPA and agrochemical industry dismissed the USDA-led study; the EPA questioned the study's "biological relevancy" to bees "under natural conditions" and cast doubt on such laboratory-based findings that suggest a link between very low levels of imidacloprid and honey bee deaths as being "highly uncertain" (EPA, 2012, p. 156). Thus, even though there are some signs of a shift in the epistemic culture of US honey bee biology toward more multifactorial laboratory-based approaches, a single-factor, field-based experimental approach remains the dominant evidentiary standard for determining the "imminent hazard" of a pesticide in the US regulatory culture (e.g., EPA, 2008a, 2012).

The primacy of a unifactorial field experimental approach to the relationship between bees and pesticides is the outcome of a particular history, an orientation that is intimately tied to the growth of entomology as a science of insecticide development in the context of state- and industry-sponsored agricultural research in the United States (Suryanarayanan & Kleinman, 2013). Consequently, while none of the involved actors would deny the possibility that imidacloprid and related chemicals are involved in recent phenomena of mass honey bee deaths, the accepted evidentiary norms of field experiments make it highly unlikely that anyone will find support for the kinds of low level, interactional effects that beekeepers have stressed based on their own field observations. Given the established history, the agrochemical industry need only point to the lack of conclusive evidence from field experiments in order to gain the higher ground in academic and policymaking circles (e.g., Bayer CropScience and Valent U.S.A. Corporation, 2012).

Relational Dynamics of Beekeepers' and Farmers' Organizations

Another important organizational axis in the struggles over restriction in the use of systemic insecticides is the relative extent to which concerned beekeepers in each country were able to mobilize material and symbolic support for their struggles among beekeepers and others, notwithstanding considerable variation in the perceptions of individual beekeepers in both places. In France, the three national-level beekeepers' unions, which represent the interests of professional beekeepers from different regions, stood together in voicing their collective opposition against Gaucho® and Régent TS®. Crucially, the French beekeepers were actively supported in their movement by several agricultural farmers' groups, many of which belonged to the most influential and prominent national-level umbrella organization, the Federation Nationale des Syndicats d'Exploitants Agricoles (FNSEA). Compared to other civil society organizations in France, agricultural interest groups such as the FNSEA enjoy significant informal access and consultative privilege with state officials at national and local levels, and they have a prominent advisory role in implementing agricultural policy (Culpepper, 1993). But the FNSEA's membership of farmers' groups was divided over the controversy. For example, the FNSEA's Maize Grower's Association (the Association Generale des Producteurs de Mas, or AGPM) was against the beekeeper's collective opposition to Gaucho® (La France Agricole, 2002). But other member groups such as the FDSEA (Fédération Départementale des Syndicats d'Exploitants Agricoles, or County Federation of Agricultural Unions) of Vendèe and the Confèdèration Paysanne (Country Confederation) were strongly in support; this support was strong enough that the FNSEA itself declared its support for the French beekeepers' movement in a national demonstration against Gaucho®, where an estimated 2,000 people gathered outside Bayer's factory in October 2000 (FNSEA, 2000). Indeed, given the varied perspectives among farmer groups, there was little prospect of a substantial mobilization by farmers against the beekeepers' movement.

By contrast, in the United States, the beekeepers' movement has confronted substantial division among prominent actors within the commercial beekeeping community. Since 1969, when the American Honey Producers Association split from the American Beekeeping Federation (Moffett, 1980), these two national-level beekeeping organizations have had tense relations and have not presented a unified front in the movement against agrochemical industry. Consequently, a group of primarily affected and powerful commercial beekeepers from both organizations took the initiative to form a new organization, NHBAB, in late 2008. The formation of NHBAB – an organization of some, but certainly not all US commercial beekeepers – reflects divisions within the US commercial beekeeping community. There are significant disagreements among commercial beekeepers, some of whom wield considerable influence among beekeepers and beyond, regarding the cause(s) of CCD, and the role of the newer systemic insecticides. While many beekeepers see systemic insecticides as major contributors to CCD and favor taking these chemicals off the market, others are more skeptical. Thus, for example, Randy Oliver, a prominent and respected member of the beekeeping industry and a voice listened to by academic scientists, regulators, and agrochemical industry actors, has consistently contended that he has "read enough studies, and spoken to enough beekeepers on the subject to realize that there is not enough hard data, or clearly verifiable anecdotal reports to support the hypothesis" that seeds treated with the newer systemic insecticides (such as Gaucho®) "cause noticeable problems" to beehives "in the field, in either long- or short-term trials" (R. Oliver, personal correspondence, November 2009). Another set of commercial beekeepers, identified by the NHBAB's beekeepers as "politically well connected," goes even further, arguing that CCD is just a case of “PPB” or “piss poor beekeeping.” For these beekeepers, the external “chemical environment” is beyond one’s control, and only beekeepers who “adapt and change” their beekeeping practices as the environment changes will continue to succeed and make profit. They consider newer systemic insecticides such as clothianidin seed-coating (Poncho®) in corn to be “the best thing that [has] ever happened for beekeepers” compared to “planes flying all day dumping on” older insecticides such as “Penncap, Furadan, or Parathion.” They caution the NHBAB to stop its efforts against the newer systemics, claiming that “there are many more beekeepers that feel the same way” as them, and that “these beekeepers will also come out of hiding to fight and stop the message real quick” (*ibid.*).

These dissenting discourses have not been marginal; rather, they figure as key factors to be resolved in “strategic retreat” meeting reports of the NHBAB, and, we believe, they impaired the NHBAB’s attempts to speak with “one voice” for the US honey bee industry. Beyond the beekeepers themselves, several influential farmers’ groups such as the National Corn Growers Association, the National Cotton Council, the National Potato Council, and the National Sunflower Association among others, and multiple state departments of agriculture stood with the agrochemical industry in lobbying the EPA to keep the market authorization of the agrochemicals intact. For example, in the “public comment” period leading up to the EPA’s eventual re-registration of imidacloprid in 2009, representatives from these national crop consortia repeatedly emphasized the crucial role played by the chemical in controlling economically significant insect pests. They also highlighted imidacloprid as the “most environmentally sound control” with no better alternatives, and suggested that there would be dire consequences for their industry and the public if the chemical were to be withdrawn from the market.[6] In this context, US government officials had no reason to rethink established evidentiary standards or the kinds of policies appropriate in the face of uncertainty about causal factors.

State Structures of Regulatory Policymaking

In France, the relatively centralized and impermeable structure of regulatory policymaking encouraged direct confrontation in the form of organized street protests by disempowered beekeeper groups, because institutional engagement with the state was especially challenging.[7] Furthermore, the policymaking structure also provided the Minister of Agriculture with the discretionary authority to overturn expert recommendations. France in the 1990s, despite the gradual transition to the more fragmentary politics of the nascent European Union, remained a largely centralized state (Guyomarch & Machin, 2001) with relatively weak involvement of civil society in policymaking (Keeler & Hall, 2001). This meant that the French beekeepers could not participate meaningfully in the debates that took place between scientists, agrochemical actors and policymakers in the Ministry of Agriculture (Prendiville, 1994). Lacking access to the consultative processes of policymaking, disempowered citizens' groups have tended historically to confront state power via organized street protests, sit-ins, marches and demonstrations. That type of civic engagement has historically gained much more tolerance, traction, and even outright support among the public than is typical in other nation-states (Keeler & Hall, 2001).

The tactics of street protests, far from being unconventional, are "politics as usual" in the organizational landscape of French politics, where it has become customary for those with grievances to seek redress from the state and for the state to respond, ironically, often sympathetically.[8] Indeed, there have been numerous occasions in the past where protests led the government to withdraw controversial proposals or give key concessions (Keeler & Hall, 2001). Seen in this light, the precautionary decisions taken by successive Ministers of Agriculture in 1999 and 2004 can be understood as a reflection of the ways in which beekeeper-organized protests prompted state officials to rethink established assumptions about the relationship between scientific uncertainty, evidentiary standards, and government policy.

On the other hand, the highly fragmented structure of the US state encourages division within and between the legislative and executive branches. Moreover, the permeability of the US state legislative branch and regulatory institutions means that civil society and interest groups have multiple points at which to make their voices heard (Kleinman, 1995). These features have shaped the dispersed form taken by beekeepers' collective actions in the United States, and this action never led to serious challenges to evidentiary norms in the US EPA.

Policymaking on pesticides in the United States was originally under the purview of the Department of Agriculture (USDA), but in 1970, under the Nixon administration, this responsibility shifted to the EPA (Jasanoff, 1990). Since then, pesticide policymaking has developed in a highly adversarial context, featuring active interventions by the Congress and the federal courts on behalf of an array of industry, labor, and environmental interests. These regulatory outcomes under the Toxic Safety Control Act of 1976 and the Food Quality Protection Act of 1996 weakened the EPA director's discretionary powers and oversight by the USDA and led to increased deference to the recommendations of scientific advisory panels and to greater consideration of economic impacts in risk assessments (Brickman et al., 1985; Jasanoff, 1990). As a result, the NHBAB's beekeepers could not direct their collective actions *and* resources against any single authority or agency, and they could not hope that the action of a single government ally – the head of the EPA – would transform their cause. Instead, their time and resources had to be spread between lobbying Congress and its many relevant committees and engaging the EPA. At the same time, efforts by labor and environmental interests in the 1970s and 1980s opened up the chemical policymaking process to limited access by interest groups and civil society through mechanisms such as public comment and the Freedom of Information Act (Brickman et al., 1985). But multiplying the points of access also allowed advantages to stakeholders with greater resources, who could intervene extensively at multiple opening points in the state. Thus, somewhat counterintuitively, agrochemical company interventions weakened the potential for beekeepers and their allies in civil society to mobilize for their claims.

CONCLUSION

We began by asking why the governments of France and the United States came to contrasting conclusions regarding scientific evidence and policy related to honey bee die-offs potentially from the same set of insecticides. On the one hand, the French government in 1999 and 2004 suspended the commercial use of the insecticidal products that beekeepers suspected of causing the honey bee declines. On the other hand, the US government has to date rejected beekeepers' calls to limit the usage of the very same set of insecticides. We argued that the different outcomes are underpinned by divergent forms of beekeepers' resistance in each place and the different contexts within which that resistance takes place. The beekeepers' collective resistance in France was a unified and sustained "social movement" replete with media campaigns, strikes and street protests, whereas in the US beekeeper engagement was characterized by a relatively dispersed set of coordinated actions. The French government, jarred by social movement pressure, developed a regulatory policy in this instance that, in the face of conflicting and less than definitive studies, was undergirded by a preference for false positive results in some of these studies and caution in the face of scientific uncertainty. We argued that the French beekeepers' position was strengthened because they could rely on suggestive results from laboratory experiments in a way that US beekeepers could not, and that different epistemic cultures of honey bee biology explain the contrasting status of the knowledge produced in laboratory and field toxicity studies. These are positions about how to treat knowledge and more particularly when to ignore or tentatively accept data that is not definitive. Without sustained pressure from beekeepers and civil society actors, a false negative error norm has remained the unshaken basis for US policy in the controversy over newer systemic insecticides and honey bee deaths in the United States. In France, the three national-level beekeepers' organizations spoke with the one voice that the NHBAB's beekeepers in the United States aspired to, but could not, achieve. The relatively centralized and impermeable structure of French regulatory policymaking not only made more likely direct confrontation in the form of organized street protests by disempowered beekeeper groups but also allowed the Minister of Agriculture the discretion to overturn expert recommendations. In the United States, the beekeepers' capacity to organize was weakened by fractious relations between prominent actors within the beekeeping industry. The decentralized structure of regulatory policymaking in the US state fragmented the beekeepers' resistance, making it harder for them to mobilize resources against a single target or authority.

Much like recent comparative historical scholarship in STS that explores the ways in which variation in state structures, epistemic cultures, the organization of civil society, and national cultures can shape outcomes of interest, we have examined how differences between the United States and France along related dimensions shaped the ways in which the two governments dealt with scientific knowledge and uncertainty and determined policy. Here, we look at the intersection of states and social movements, attending specifically to how national historical legacies shape state and social movement capacities and how the quite different social organization of state-social movement nexus in France and the United States produced different orientations to scientific data and divergent (temporary) resolutions to the controversy over the massive deaths of honey bees and the role of newer systemic insecticides in contributing to the bees' demise.

It has become common wisdom in STS that data alone rarely resolve technoscientific controversies (e.g., Collins, 1985). The dispute we investigated is consistent with this view, but like so many of the crucial science-related conflicts of the day, it is not negotiation among scientists on an even playing field that produces closure. Our work points to the importance of exploring the role of nonscientist stakeholders in technoscientific controversies in the public realm and the ways in which the capacities of different stakeholders are shaped by the social organization of national polities and cultures of knowledge making. More specifically, we hope our work will encourage further research that looks at the intersection of states and social movements in technoscientific controversies and takes seriously the ways in which national variation in the social organization of the state-social movement nexus shapes the capacities – indeed, the power – of relevant stakeholders and thus the contours of the resolution of such disputes. As technoscientific controversies increasingly have direct impacts on citizens across the globe, it strikes us that this is a crucial realm for investigation by STS scholars.

NOTES

1. Imidacloprid represents a newer generation of "systemic" insecticides, which are extremely potent insect neurotoxins whose chemical properties enable them to persist in treated plants, and thus supposedly enable reduced quantities of pesticides to carry out targeted pest control for longer durations (Elbert et al., 1990).

2. For example, Colin et al. (2004) simulated "field" conditions by constraining artificially small beehives to "insect-proof" tunnels that were 4 m high and 8 m × 20 m in surface area, where the bees were "trained" to forage on sugar syrup mixed with specific doses of systemic insecticides.

3. European Commission Decision No. 485, 2013, *Official Journal of the European Union* L 139/12.

4. In contrast to the EPA's regulatory role, the USDA has been more involved in coordination and implementation of scientific research and communication about the spate of honey bee deaths (e.g., USDA, 2010).

5. "Field experiments" are considered by bee scientists to be distinct from "monitoring" studies, which measure aspects such as pesticide residues in various field settings (Maxim & van der Sluijs, 2013).

6. Letter from John Keeling, Vice President of the National Potato Council to the EPA. EPA Docket # EPA–HQ–OPP–2008-0844 (March 17, 2009); Letter from E. Keith Menchey, National Cotton Council of America, EPA Docket #EPA–HQ–OPP–2008-0844 (March 16, 2009); Letter from Richard Gaskalla, Florida State Department of Agriculture, EPA Docket # EPA–HQ–OPP–2008-0844 (March 12, 2009).

7. The idea that a centralized and relatively impermeable regulatory structure encourages street protest complicates previous broad characterizations of regulatory policymaking among Western European nation-states as being "cooperative" (e.g., Jasanoff, 1986).

8. This is not to make any generalization that the US is not a "protest society" and that France is one. There is little empirical support for the popular imagery of France being prone to more protests than other nation-states (Wilson, 1987).

ACKNOWLEDGMENTS

We are grateful to Luc Gomel (Micropolis, Aveyron, France) for pointing us to important sources of information regarding the French controversy. We also thank David J. Hess and Scott Frickel for suggesting ways to improve our manuscript. This work was supported by the National Science Foundation (Award # 0924346). Daniel Lee Kleinman's work is also supported by the National Research Foundation of Korea Grant (NRF-2010-330-B00169).

REFERENCES

Bayer CropScience and Valent U.S.A Corporation. (2012, September 24). Response to the Clothianidin Petition: Attachment 1 – Technical Response. Docket # EPA-HQ-OPP-2012-0334-1324. Retrieved from http://www.regulations.gov/#!documentDetail;D=EPA-HQ-OPP-2012-0334-1324. Accessed on December 31, 2013.

Belzunces, L., & Tasei, J.-N. (1997). *Rapport sur les effets des traitements de semences de tournesol au Gaucho® (imidaclopride). Impacts sur les peuplements des colonies d'abeilles et sur les miellées* (p. 27). Paris: Commission d'Etude de la Toxicité des Produits Antiparasitaires à Usage Agricole et des Produits assimilés, Ministère de l'Agriculture.

Blacquière, T., Smagghe, G., van Gestel, C. A. M., & Mommaerts, V. (2012). Neonicotinoids in bees: A review on concentrations, side-effects and risk assessment. *Ecotoxicology*, *21*, 973–992.

Bonmatin, J.-M., Moineau, I., Charvet, R., Colin, M. E., Flèchè, C., & Bengsch, E. R. (2003). A sensitive LC/APCI/MS/MS method for analysis of imidacloprid in soils, in plants and in pollens. *Analytical Chemistry*, *75*, 2027–2033.

Brickman, R., Jasanoff, S., & Ilgen, T. (1985). *Controlling chemicals: The politics of regulation in Europe and the United States*. Ithaca, NY: Cornell University Press.

Burgett, M., Daberkow, S., Rucker, R., & Thurman, W. (2010). U.S. pollination markets: Recent changes and historical perspective. *American Bee Journal*, *150*, 35–41.

Chauzat, M.-P., Cauquil, L., Roy, L., Franco, S., Hendrikx, P., & Ribière-Chabert, M. (2013). Demographics of the European apicultural industry. *PLoS ONE*, *8*(11), e79018.

Colin, M. E., Bonmatim, J. M., Moineau, I., Gaimon, C., Brun, S., & Vermandere, J. P. (2004). A method to quantify and analyze the foraging activity of honey bees: Relevance to the sublethal effects induced by systemic insecticides. *Archives of Environmental Contamination and Toxicology*, *47*, 387–395.

Collins, H. (1985). *Changing order: Replication and induction in scientific practice*. Chicago, IL: University of Chicago Press.

Confèdèration Paysanne. (2003, April). Fipronil: Retrait Immediate. La Confédération Paysanne Invite Les Paysans Au Boycott. Retrieved from http://www.beekeeping.com/intoxications/communique_conf_04_08_03.htm. Accessed on February 8, 2012.

Coordination des Apiculteurs de France. (1998). Manifestation de Paris. Retrieved from http://www.apiservices.com/articles/fr/com_gaucho.htm

CST (Scientific and Technical Committee for the Multifactor Study of the Honeybee Apiaries Decline). (2003). Imidaclopride Utilisé en Enrobage de Semences (Gaucho®) et Troubles des Abeilles, Rapport Final. Paris: Ministère de l'Agriculture, de la Pêche et des Affaires Rurales.

Culpepper, P. D. (1993). Organisational competition and the neo-corporatist fallacy in French agriculture. *West European Politics*, *16*(3), 293–315.

Daberkow, S., Korb, P., & Hoff, F. (2009). The structure of the US beekeeping industry. *Journal of Economic Entomology*, *102*(3), 868–886.

Decourtye, A., Le Metayer, M., Pottiau, H., Tisseur, M., Odoux, J. F., & Pham-Delegue, M. H. (2001). Impairment of olfactory learning performances in the honey bee after long term ingestion of imidacloprid. In L. P. Belzunces, C. Pelissier, & G. B. Lewis (Eds.), *Hazards of pesticides to bees* (pp. 259–268). Paris, France: INRA.

Desneux, N., Decourtye, A., & Delpuech, J.-M. (2007). The sublethal effects of pesticides on beneficial arthropods. *Annual Review of Entomology*, *52*, 81–106.

Elbert, A., Becker, B., Hartwig, J., & Erdelen, C. (1990). Imidacloprid: A new systemic insecticide. *Pflanzenschutz-Nachrichten Bayer*, *44*, 113–136.

EPA (US Environmental Protection Agency).(2008a, October 10). *EPA response to Sierra Club's request to suspend nicotinyl insecticides.* Docket # EPA-HQ-OPP-2008-0844-0120.

EPA (US Environmental Protection Agency).(2008b, October 7). Pollinator protection in the EPA's Office of pesticide programs. *Pesticide program dialogue committee meeting transcript* (pp. 239–286). Retrieved from http://www.epa.gov/oppfead1/cb/ppdc/2008/oct2008/transcript.pdf. Accessed on October 26, 2009.

EPA (US Environmental Protection Agency). (2010, October 15). Notice of Conditional Pesticide Registration for Movento. EPA Registration No. 264–1050.

EPA (US Environmental Protection Agency). (2012, July 17). Technical support document for the response to the emergency citizen petition seeking suspension of registration for clothianidin based on claims of imminent hazard to the environment. Docket # EPA-HQ-OPP-2012-0334. Retrieved from http://www.regulations.gov/#!documentDetail;D=EPA-HQ-OPP-2012-0334-0012. Accessed on July 19, 2012.

Fischer, D., & Moriarty, T. (2011). *Pesticide risk assessment for pollinators: Summary of a SETAC Pellston Workshop*. Pensacola, FL: Society of Environmental Toxicology and Chemistry (SETAC) Press.

FNSEA (Federation Nationale des Syndicats d'Exploitants Agricoles). (2000). *Press release: Save the beehives*. Retrieved from http://www.apiservices.com/articles/fr/gaucho/manifestation_cormery.htm. Accessed on February 8, 2012.

Fourcade, M. (2009). *Economists and societies: Discipline and profession in the United State, Britain, and France, 1890s to 1990s*. Princeton, NJ: Princeton University Press.

Frickel, S., Gibbon, S., Howard, J., Kempner, J., Ottinger, G., & Hess, D. J. (2010). Undone science: Charting social movement and civil society challenges to research agenda setting. *Science, Technology, & Human Values*, *35*(4), 444–473.

Frickel, S., & Moore, K. (2008). *The new political sociology of science*. Madison, WI: The University of Wisconsin Press.

Gruzelle, F. (2011, October 1). 15 Octobre 2011: Manifestations Pour La Défense Des Abeilles. Retrieved from: http://suite101.fr/article/15-octobre-2011-manifestations-pour-la-defense-des-abeilles-a31646. Accessed on May 3, 2013.

Guyomarch, A., & Machin, H. (2001). Political developments in their historical institutional context. In A. Guyomarch, H. Machin, P. A. Hall, & J. Hayward (Eds.), *Developments in French politics 2* (pp. 1–22). London: Palgrave.

Halm, M.-P., Rortais, A., Arnold, G., Tasei, J. N., & Rault, S. (2006). New risk assessment approach for systemic insecticides: The case of honey bees and imidacloprid (Gaucho). *Environmental Science and Technology*, *40*, 2448–2454.

Henry, M., Bèguin, M., Requier, F., Rollin, O., Odoux, J.-F., Aupinel, P., … Decourtye, A. (2012). A common pesticide decreases foraging success and survival in honey bees. *Science*, *336*(6079), 348–350.

Jasanoff, S. (1986). Risk management and political culture: A comparative study of science in the policy context. *Social research perspectives: Occasional reports on current topics* (No. 12). New York, NY: Russell Sage Foundation.

Jasanoff, S. (1990). *The fifth branch: Science advisors as policymakers*. Cambridge, MA: Harvard University Press.

Jasanoff, S. (2005). *Designs on nature: Science and democracy in Europe and the United States*. Princeton, NJ: Princeton University Press.

Johnson, R. M., Ellis, M. D., Mullin, C. A., & Frazier, M. (2010). Pesticides and honey bee toxicity – USA. *Apidologie*, *41*(3), 312–331.

Keeler, J. T. S., & Hall, P. A. (2001). Interest representation and the politics of protest. In A. Guyomarch, H. Machin, P. A. Hall, & J. Hayward (Eds.), *Developments in French politics 2* (pp. 50–67). London: Palgrave.

Kinchy, A. (2012). *Seeds, science, and struggle: The global politics of transgenic crops*. Cambridge, MA: MIT Press.

Kitschelt, H. P. (1986). Political opportunity structures and political protest: Anti-nuclear movements in four democracies. *British Journal of Political Science*, *16*, 57–85.

Kleinman, D. L. (1995). *Politics on the endless frontier: Postwar research policy in the United States*. Durham, NC: Duke University Press.

Kleinman, D. L. (2003). *Impure cultures: University biology and the world of commerce*. Madison, WI: University of Wisconsin Press.

Kleinman, D. L., & Kinchy, A. (2003). Why ban bovine growth hormone? Science, social welfare, and the divergent biotech policy landscapes in Europe and the United States. *Science as Culture*, *12*(3), 375–414.

Kleinman, D. L., & Suryanarayanan, S. (2013). Dying bees and the social production of ignorance. *Science, Technology, and Human Values*, *38*(4), 492–517.

Knorr-Cetina, K. (1999). *Epistemic cultures: How the sciences make knowledge*. Cambridge, MA: Harvard University Press.

La Coordination des Apiculteurs de France. (2000a). *A stricken sector of production in apiculture: Beekeeping, hit in full flight*. Retrieved from http://www.beekeeping.com/articles/us/gaucho/gaucho_france.htm. Accessed on February 6, 2012.

La Coordination des Apiculteurs de France. (2000b). *Composite document of present position relating to Gaucho, sunflower and bees*. Retrieved from http://www.beekeeping.com/articles/us/gaucho/manifestation_paris_us.htm. Accessed on February 6, 2012.

La Coordination des Apiculteurs de France. (2002). *Evaluation of fipronil when compared with imidacloprid*. Retrieved from http://www.beekeeping.com/articles/fr/fipronil_imidaclopride.htm. Accessed on February 6, 2012.

La France Agricole. (2002, December 13). AGPM et Gaucho (No. 2964). Retrieved from http://www.lafranceagricole.fr/Arcbeehives/articlexml/11919. Accessed on April 8, 2012.

MacIntyre, B. (1998). "Mad bee" disease hits French honey supply. *The Times*, April 9.

Maus, C., Curé, G., & Schmuck, R. (2003). Safety of imidacloprid seed dressings to honey bees: A comprehensive review and compilation of the current state of knowledge. *Bulletin of Insectology*, *56*, 51–57.

Maxim, L., & van der Sluijs, J. P. (2007). Uncertainty: Cause or effect of stakeholders' debates? Analysis of a case study: The risk for honeybees of the insecticide Gaucho®. *Science of the Total Environment*, *36*, 1–17.

Maxim, L., & van der Sluijs, J. P. (2013). Seed-dressing systemic insecticides and honeybees. In D. Gee, P. Grandjean, S. F. Hansen, S. Hove, M. MacGarvin, J. Martin, G. Nielsen, D. Quist, & D. Stanners (Eds.), *Late lessons from early warnings: Science, precaution, innovation* (Vol. 2, pp. 401–438). European Environment Agency (Report No. 1).

Ministère de l'Agriculture. (2001, February 2). *Press release*. Retrieved from http://www.beekeeping.com/intoxications/communique_ma_02_02_2001.htm. Accessed on February 7, 2012.

Ministère de l'Agriculture. (2003, October). Note de la Direction Générale de l'alimentation à l'attention des Coprésidents du Comité Scientifique et Technique. Retrieved from http://www.beekeeping.com/intoxications/communique_ma_31_10_2003.htm. Accessed on February 8, 2012.

Ministère de l'Agriculture. (2004, February 9). *Press release*. Retrieved from http://www.beekeeping.com/intoxications/communique_ma_09_02_2004.htm. Accessed on February 8, 2012.

Moffett, J. (1980). History of national associations. In *Beekeeping in the United States* (Agricultural Handbook No. 335, pp. 175–179). Washington, DC: USDA.

Moore, K. (2008). *Disrupting science: Social movements, American scientists, and the politics of the military, 1945–1975*. Princeton, NJ: Princeton University Press.

National Honey Bee Advisory Board. (2009a, March 17). Letter to the office of pesticide programs. US Environmental Protection Agency. Docket # EPA-HQ-OPP-2008-0844-0010.

National Honey Bee Advisory Board. (2009b). *A strategic planning retreat for the National Honey Bee Advisory Board*. The meeting was held on October 26–27, 2009.

National Honey Bee Advisory Board. (2011). Pollinators and pesticides 2011: Fundamental flaws of pesticide policy in the United States. Opinion paper of the National Honey Bee Advisory Board distributed at the 2011 North American Beekeeping Conference, Galveston, TX.

National Research Council. (2007). *Status of pollinators in North America*. Washington, DC: The National Academies Press.

Oliver, R. (2011). Forum: Response to Suryanarayanan and Kleinman's disappearing bees and reluctant regulators. *Issues in Science and Technology*. Retrieved from http://www.issues.org/28.1/forum.html. Accessed on April 8, 2012.

Oreskes, N., & Conway, E. M. (2010). *Merchants of doubt: How a handful of scientists obscured the truth on issues from tobacco smoke to global warming*. New York, NY: Bloomsbury Press.

Pettis, J. S., vanEngelsdorp, D., Johnson, J., & Dively, G. (2012). Pesticide exposure in honey bees results in increased levels of the gut pathogen Nosema. *Naturwissenschaften*, *99*, 153–158.

Prendiville, B. (1994). *Environmental politics in France*. Boulder, CO: Westview Press.

Ratnieks, F. L. W., & Carreck, N. L. (2010). Clarity on honey bee collapse? *Science*, *327*, 151–152.
Schmuck, R. (1999). No causal relationship between Gaucho® seed dressing in sunflowers and the French bee symptoms. *Pflanzenschutz Nachrichten Bayer*, *52*, 257–299.
Schweber, L. (2006). *Disciplining statistics: Demography and vital statistics in France and England, 1830–1885*. Durham, NC: Duke University Press.
Spivak, M. (2010). The status of the European honey bee in the US. In E. Mader, M. Spivak, & E. Evans (Eds.), *Managing alternative pollinators: A handbook for growers, beekeepers and conservationists* (pp. 15–24). Ithaca, NY: Natural Resource, Agriculture and Engineering Service.
Suchail, S., Guez, D., & Belzunces, L. (2001). Discrepancy between acute and chronic toxicity induced by imidacloprid and its metabolites in *Apis mellifera*. *Environmental Toxicology and Chemistry*, *20*, 2482–2486.
Suryanarayanan, S., & Kleinman, D. L. (2011). Disappearing bees and reluctant regulators. *Issues in Science and Technology*, *27*(4), 31–36.
Suryanarayanan, S., & Kleinman, D. L. (2013). Be(e)coming experts: The controversy of insecticides in the honey bee colony collapse disorder. *Social Studies of Science*, *43*(2), 215–240.
Tarrow, S. (2011). *Power in movement: Social movements and contentious politics*. New York, NY: Cambridge University Press.
Traweek, S. (1992). *Beamtimes and lifetimes*. Cambridge, MA: Harvard University Press.
UNAF (Union Nationale de l'Apiculture Française). (2003, September 19). Les scientifiques du Comité Scientifique et Technique confirment la toxicité du "Gaucho". Retrieved from http://www.beekeeping.com/intoxications/communique_unaf_19_09_2003.htm. Accessed on February 8, 2012.
UNAF (Union Nationale de l'Apiculture Française). (2004a, March 31). Affaire Gaucho au Conseil d'etat. Victoire des Apiculteurs! Retrieved from http://www.beekeeping.com/intoxications/communique_unaf_31_03_2004.htm. Accessed on February 8, 2012.
UNAF (Union Nationale de l'Apiculture Française). (2004b, May 25). L'interdiction du Gaucho: Une Grande Victoire Pour l'apiculture Française et Européenne! Retrieved from http://www.beekeeping.com/intoxications/communique_unaf_25_05_2004.htm. Accessed on February 8, 2012.
UNAF (Union Nationale de l'Apiculture Française). (2011, January 24). The National Union of French Beekeeping (UNAF) today announced victory at the Council of State for France, against authorization of the pesticide CRUISER®. Retrieved from http://www.unaf-apiculture.info/presse/2011_English_UNAF_Victory_announcement_over_Crusier_in_Court.pdf. Accessed on April 28, 2013.
USDA (US Department of Agriculture). (2007). Colony collapse disorder action plan by CCD Steering Committee. Retrieved from http://www.ars.usda.gov/is/br/ccd/ccd_actionplan.pdf. Accessed on July 15, 2008.
USDA (US Department of Agriculture). (2010). Colony collapse disorder progress report by CCD Steering Committee. Retrieved from http://www.ars.usda.gov/is/br/ccd/ccdprogressreport2010.pdf. Accessed on December 17, 2010.
U.S. District Court, New York. (2010). Natural Resources Defense Council and the Xerces Society v. US EPA and Bayer CropScience. LP 09, Civ. 4317.
U.S. District Court, Pennsylvania. (2008). Bauer v. Bayer AG, 564 F. Supp. 2d 365.

van Engelsdorp Dennis, J., Hayes, R. M., & Pettis, U. J. (2008). A survey of honey bee colony losses in the US, Fall 2007 to Spring 2008. *PLoS ONE*, *3*(12), e4071. doi:10.1371/journal.pone.0004071

Wilson, F. L. (1987). *Interest-group politics in France*. New York, NY: Cambridge University Press.

Wright, S. (1994). *Molecular politics: Developing American and British regulatory policy for genetic engineering, 1972–1982*. Chicago, IL: University of Chicago Press.

WHEN GREEN BECAME BLUE: EPISTEMIC RIFT AND THE CORRALLING OF CLIMATE SCIENCE

David J. Hess

ABSTRACT

The emergence of climate science denialism in the United States provides a challenge to STS theories of the relationship between scientific expertise and public policy because a situation of epistemic rift occurs: the capacity of scientific consensus to establish the grounds of political debate is broken, and the standard circulation of expertise from the scientists and funding from the state is interrupted. Three mechanisms for the containment of scientific expertise are studied: direct intellectual suppression of climate scientists, industry support of contrarian scientists and policymakers, and cutbacks on government research programs that support climate change. This situation politicizes climate scientists, who are drawn into the public sphere as a counterpublic to the effort to contain the circulation of their knowledge in the political field. Although the strategy of contained expertise has been effective in blocking climate legislation at the federal government level in the United States, it may be

Fields of Knowledge: Science, Politics and Publics in the Neoliberal Age
Political Power and Social Theory, Volume 27, 123–153
Copyright © 2014 by Emerald Group Publishing Limited
All rights of reproduction in any form reserved
ISSN: 0198-8719/doi:10.1108/S0198-871920140000027012

losing effectiveness, and an emergent alternative strategy based on adaptation may be coming to replace it. Factors that affect the reduction in the capacity to contain the circulation of scientific expertise are also analyzed.

In the United States a well-orchestrated public relations machinery has emerged to contain the policy implications of scientific research that has identified threats posed by anthropogenic climate change and the need for a policy response (Dunlap & McCright, 2011; Jacques, Dunlap, & Freeman, 2008; Oreskes & Conway, 2010). The machinery subjects legitimate climate scientists to personal attacks by industry-funded scientists and conservative media, and it enables elected politicians to voice climate denialism and skepticism without danger of an immediate collapse of credibility among voters.[1] For example, during the 2012 presidential campaign, candidates for the Republican nomination for president expressed anti-science views with respect to climate change, often in contrast to views that they had formerly professed to hold.

The emergence of climate science denialism among political leaders presents a puzzle for theories of scientific expertise and the policy process, because there has been a partial breakdown in the circulation of expertise from the research community to the political field and in the funding of the research community by the government. This study uses the case of the partial breakdown in credibility of an expert community as a basis for building field theory on political power and the relations between the scientific field and other social fields, such as the state, industry, and civil society. Although Bourdieu (2001) produced an important set of theoretical concepts for understanding change within the scientific field, he did not develop a general theory of the relations between the scientific and other social fields (Camic, 2011). Several approaches have emerged and are discussed in this volume. One approach, which draws directly on Bourdieu's work, is to focus on the changing relationship between the basic and applied poles of research fields (Albert & McGuire, this volume; Lave, 2012) that in turn is an outcome of broader societal and political changes. Another approach is to examine how changes in other social fields alter the political opportunity structure to permit challengers to contest the positions of incumbents, an approach that might include the effects of political funding structures on strategies of conservation and subversion in the scientific field (Fligstein & McAdam, 2012). Yet another approach is to examine how policy changes affect the relative position of different institutional logics, logics of action, or patterns of governmentality in the scientific field (Berman & Schweber, this volume).

This study builds on the conversation in the political sociology of science on interfield relations by moving in a somewhat different direction based on the problem posed by climate denialism, that is, the problem of how and why an established interfield relationship breaks down. Thus, the goal is not to explain changes within the scientific field by reference to an external change, such as the shift of policymakers toward a greater concern with innovation and with neoliberal policy instruments. Rather, the goal is to begin with an established relationship between the scientific and political fields and to explain how the relationship has broken down.

The relationship between the scientific field and other social fields can be viewed as a kind of metabolism or exchange relationship. The scientific field produces knowledge for the other social fields, and in turn it receives both support and a degree of autonomy to pursue the manufacture of knowledge. Support can include direct, financial support in the form of contracts and grants, and it can also include broader support from the private sector and civil society for government funding of the research and educational enterprises. However, the exchange relationships can break down, and the term "epistemic rift" is introduced here based on the similar concept of "ecological rift" in environmental sociology, which was developed from the study of the relationship between agriculture and human waste that breaks down with the urbanization process. Whereas in traditional agrarian societies agricultural and animal waste is returned to the soil, in urban societies the metabolism breaks down, and the human waste is disposed in landfills and water resources (Foster, 2000). In a similar way, an "epistemic rift" can occur when significant segments of the state (or other social fields) reject the credibility of scientific research and withdraw their support for it.

This study will develop some of the theoretical implications of the epistemic rift based on a case study of climate denialism in the United States. The first section will outline in more detail a conceptual toolkit for understanding epistemic rift, the second section will discuss the historical background of climate denialism in the United States, the third section will delineate the mechanisms of epistemic rift, and the fourth section will discuss the emergence of a possible resolution to the crisis and the causes for the shift. The study is largely theoretical and in the Weberian tradition of social theory. In other words, it aims at developing typological mechanisms of the dynamics of epistemic rift that can elucidate empirical research. In the process, the project draws on the research base of two long-term ethnographic projects on the green economy in the United States, with two teams of student researchers who conducted semi-structured interviews with state and local leaders in the government, business, and nonprofit sectors.

CONCEPTUAL BACKGROUND

This research project draws on and contributes to one theoretical current within the political sociology of science and technology, field theory, by focusing on the issue of interfield relationships. The term "field" is used here to describe a space of social contestation where the actors have a common stake in outcomes, such as a policy field in which there are relations of cooperation and conflict over policymaking and implementation (Bourdieu, 2005; Fligstein & McAdam, 2012). In contrast with the focus on norms, processes, and rules found in traditional functionalist analyses of institutions, the field concept is preferred here because of its greater emphasis on power, strategies of conflict and cooperation, and the diversity of agency. The approach to field theory adopted here also pays attention to the role of systems of meaning, but it utilizes the concept of field-based cultural systems rather than the more individualized concept of habitus. Specifically, the analysis will focus on the role of clashing political ideologies such as neoliberalism, social liberalism, and developmentalism in the political field. In the case to be discussed below, longstanding ideological conflicts between neoliberalism and social liberalism over the regulatory and redistributive role of the state are central.

Power and State Theory

The political sociology of the modern, parliamentary state has generally found a poor empirical match between observed political processes and two, opposing ideal types of the state: an instrument of the interests of a dominant economic class and/or a political elite, and an independent arbiter among relatively coeval interest groups mobilized for specific issues. Instead, there is broad recognition of the variation across policy issues, in which some issue networks show domination by economic and political elites and others are closer to a pluralist model (Hicks & Lechner, 2005;

McFarland, 2010). Although actors with high levels of financial resources can strongly influence both public opinion and policy outcomes for issues that they deem important, both economic elites and political leaders are often divided. Consequently, policy outcomes are often best explained by a focus on coalitions of actors that emerge and dissolve around specific issues. Rather than reject theories of the state based on class, political elites, or pluralist interest groups, a coalition-politics approach to a conflict theory of the state brings those approaches together in one framework (van den Berg & Janoski, 2005).

Power is understood here as the capacity of an agent to influence the course of events in a field. As Lukes (1974) has argued, there are various faces or dimensions of power, including more covert forms such as the capacity to influence the agendas and the goals of other agents. As Foucault (2007, 2008) and Rose (1999) have also argued, the more covert forms of power can be embedded in organizational routines and everyday habits. But even power in these more covert forms is connected to questions of cui bono; thus, from a conflict theory perspective, a full analysis of power must be anchored in conflict among agents, even if the agents are collective ones such as races, classes, and genders. The capacity of an actor to influence the outcome of political conflict depends on control over different types of capital, which can be imported from other social fields. Among the most important types of capital in the political field are financial (the resources needed to support lobbying and campaign activity), social and cultural (coalition partners and their knowledge of the political system), general symbolic (media credibility and public legitimacy), and epistemic (knowledge about an issue, including scientific knowledge). Because the political field has some autonomy from other social fields, the exercise of power in the political field must translate extrafield resources into the currency of the political field, which in electoral democracies means the capacity to influence votes in legislatures and implementation decisions in the executive branch (Bourdieu, 2005).

Science and Interfield Relations

Like the political field, other social fields have their own way of assessing field position and a common understanding of what is at stake. In the industrial field, the stakes are about preserving and growing an economic position, which can include a firm's position as well as an industry's position (e.g., fossil fuels) within a broader industrial sector (e.g., energy).

Thus, "market position" is the common understanding of what is at stake. In the scientific field (which is divided into thousands of research fields), there is conflict and cooperation over creating knowledge that one or more research fields deems important. What is "at stake" is the capacity to produce knowledge that the field recognizes as important and rewards with citations, prizes, and other forms of recognition (Bourdieu, 2001).

In a standard relationship of metabolism between the scientific field and other social fields, the scientific field exerts a high level of influence over the other fields because of its capacity to establish "doxa," that is, the underlying area of the uncontested that is created through the engagement of orthodoxies and heterodoxies. In the political field, when both parties accept a body of scientific expertise, there is a doxa of what *is* the case upon which political contestation is grounded. Of course, other types of knowledge may be included in policy deliberations, such as industrial and occupational expertise and the perspectives of lay people (e.g., Irwin, 2006). However, the scientific field has paramount legitimacy as the site where competing truth claims can be vetted and evaluated. This truth-determining capacity limits but does not determine the myriad battles over what *should be* the case, that is, the normative debates about how to respond to what is the case. In the standard model by which the scientific field and the political field are connected, policymakers draw on scientific experts to provide advice on technical issues, and scientists engage in various strategies to create boundaries between their technical expertise and the normative claims at play in the policy field (e.g., Guston, 2001; Jasanoff, 1994). By creating a boundary between scientific expertise and policy contestation, scientists protect their doxological power at the expense of having effective political power over policy outcomes.

This standard model is also normative in the sense that it represents an ideal of the relationship between the scientific and political field that rests on a division of labor between "is" and "ought." Scientists provide knowledge about what is the case, what problems need to be addressed, and what would likely happen in a particular policy scenario, but political actors negotiate social and value conflicts to determine a policy outcome. Certainly, scientists can and do make normative statements, but the normative statements tend to be limited in scope (e.g., there is a problem that needs to be addressed). Scientists tend to separate their political positions from institutionalized advising mechanisms, which rest on a precarious defense of a perception of political neutrality.

This standard relationship may appear to be innocent of power differentials, especially to those in the main circuits of the flows of capital and knowledge. In other words, there are power differentials within the political field, but scientists have a process of reaching a high enough level of consensus on an issue to allow them to address the epistemic needs of policy-makers. However, there are often networks located in the subordinate positions of the scientific field that do not agree with the dominant research agendas and the associated definitions of public good and policy direction. In general, those in dominant positions in the scientific field reject the epistemic challenges that emerge within the field. The dominant networks of the scientific field can then mobilize various types of capital to result in the marginalization of challengers, loss of funding for them, and social exclusion.

When these conflicts remain internal to the scientific field, there is simply a situation of scientific controversy. However, in some cases those in the heterodox position in the scientific field also believe that their alternative research agenda has broad policy implications and represents a better formulation of public benefit that existing research agendas. Examples have been studied in detail in the health field, but there are also examples of "scientific and intellectual movements" more generally (e.g., Brown, 2007; Frickel & Gross, 2005; Hess, 2007, 2014). In addition to mounting an epistemic challenge within the research field, scientists can become associated with broader social movements that advocate for alternative technologies and an end to some aspects of an existing industrial regime. When scientists adopt a public position by arguing that an alternative research program would be of broad public benefit, in contrast with the existing dominant paradigm, and when they become connected with broader agendas for policy change, they form a scientific counterpublic (Hess, 2011). They also point to a condition of undone science, that is, the systematic underfunding of research programs that they argue are potentially of broad public benefit (Frickel et al., 2010; Hess, 2007). They may lose funding for their research programs and access to prestigious journal venues, but in some cases scientists can find support from large civil society organizations and firms in a countervailing industry (such as nutritional supplements companies that can counterbalance a lack of interest in complementary medicine from the pharmaceutical industry; Hess, 2009). The long-term outcome may be the incorporation and transformation of the alternative technologies into existing industrial regimes and the development of funding and recognition for the alternative research agendas within the scientific field.

In the mobilizations of subordinate networks as scientific counterpublics, there is not necessarily an epistemic rift in the general sense of a breakdown of metabolism between the scientific and political fields. The circulation of research and funding in the dominant networks of both the scientific and political fields can continue to flow, and the dominant networks of both fields can engage in various strategies to handle the challengers: ignore them, marginalize them, or incorporate and transform them. Indeed, industry often prefers to have a scientized and technocratic policy field, because the production of scientific knowledge is resource intensive, and social movements and interest groups that mount challenges to industry in the policy arena generally lack the ability to produce countervailing knowledge (Kinchy, 2012). However, when there are enough scientists who challenge the dominant paradigms of the research field, they can throw the research field into a high level of controversy. In response to the threat of a breakdown in credibility of the dominant paradigms, the dominant networks of the research and policy fields may respond by creating new flows of funding that address the undone science. In the United States a good example is the creation of the National Center for Complementary and Alternative Medicine, which responded to the challenge of complementary and alternative researchers, doctors, and patient advocacy groups by creating (and controlling) funding for the alternative research field (Hess, 2014). Thus, the funding for the mainstream programs of cancer research and their credibility in the political field were not disrupted, but new avenues of credibility and funding were opened up for the alternative programs.

The emergence of climate denialism is different because there is a breakdown of general credibility for a research field that is in a state of relatively high consensus with respect to empirical questions such as the role of anthropogenic greenhouse gases in climate change. Although there are controversies within the field, on the policy-relevant scientific issues (e.g., do anthropogenic greenhouse gases cause global warming?) there is a high level of consensus. The focus of conflict is less within the scientific field between those who support a dominant paradigm and those who oppose it than it is between industries that are producing a technology that would be harmed by the new policy direction (e.g., the fossil-fuel industries) and the networks of scientists and public-interest advocates who have identified a policy problem and solution (e.g., reduce greenhouse gas emissions). In this situation the scientific research field provides knowledge about a policy problem that is in conflict with an industry that is responsible for creating the problem, and the research field plays the role of a countervailing power to industrial power in the policy field. Industry responds to the challenge by funding contrarian science, that is, expertise that attacks the countervailing science and/or creates new empirical research (Oreskes & Conway, 2010). The goal of contrarian science is to create a false perception in the political field that the research field is in a state of high controversy and therefore to break down the doxological capacity of the scientific field to define the epistemic grounds of policy action. In the case of climate science in the United States, the contrarian science strategy is not effective at creating high levels of controversy within the research field, because most of the contrarian scientists lack standing (symbolic capital and sometimes cultural capital) *within* in the scientific field. However, the contrarian strategy is effective at causing epistemic rift because funding from a network of industry and conservative donors provides a platform for disseminating disinformation about the level of scientific controversy in the media and for funding political candidates who deny the scientific research and seek to defund it.

The epistemic rift is not general in the political field. In the United States after 2009, it became widespread among many Republicans leaders but less so for Democrats. However, its effect is to allow political leaders who are aligned with the fossil-fuel industry to portray climate scientists as politically motivated. Thus, their doxological capacity is delegitimated, and to the extent that they have an effect in the political field, they become portrayed as part of the countervailing counterpublic of environmentalists and progressives who seek broader policy reforms (see Table 1).

In summary, for the case of climate science in the United States after the 2010 elections, when denialist political leaders gained power in the US House of Representatives and some state governments, there is a need to extend theory on the relationships among the scientific, political, and industrial fields by exploring the transformation of a research community from a standard model of advice giving into a countervailing counterpublic. In this situation climate scientists have lost enough credibility among enough leaders in the political field that their expertise can no longer define the grounds of political debate.

THE HISTORICAL AND POLITICAL CONTEXT OF EPISTEMIC RIFT

From a field sociological perspective, the development of the epistemic rift for climate science was enabled by the slow transformation of the organizational field that Barley (2010) has described as "corralling." His focus and mine is on the United States, but to some extent a similar change has occurred in other areas of the world, such as the European Union

Table 1. Three Configurations of Interfield Relations for the Scientific Field.

Type of Interfield Relationship	Scientific Field	Policy Field	Industrial Field	Civil Society Field
Standard expertise	Dissensus in the research field is managed through boundary organizations	Balances advice from scientists, industry, and civil society organizations	Affects the policy process through lobbying, election participation, etc.	Affects the policy process through lobbying, election participation, etc.
Alternative scientific counterpublic	Scientists in subordinate positions openly question the dominant networks of the research field; identify undone science	Ignores calls for funding undone science unless there is a strong public mobilization and/or countervailing industry support	Ignores calls for funding, but advocates may find a nearby industry (countervailing) that supports the research	Identifies areas of undone science, recruits allies in the political and industrial fields
Countervailing scientific counterpublic	Problems identified by scientists but denied by industry; relatively low controversy among scientists not funded by industry	A battleground between countervailing science and an industry	Sunset industry funds contrarian scientists and corrals scientists; countervailing industry may emerge	Divided between industry-funded groups and grassroots mobilizations

(Greenwood, 2002; Mahoney, 2008). The "corralling" of a government occurs when large corporations build an institutional field around a government to limit the redistributive and regulatory policies associated with the ideology of social liberalism. The transformation of political ideology in the United States was based on corporate spending on political action committees, public relations firms, media with a free-market bias, conservative think tanks, lobbying firms, trade associations, and public affairs offices. This general transformation of the organizational field of the American federal government made it possible for neoliberal ideology to flourish even when Democrats were in power, because the center of political debate shifted to the right. I extend this approach by showing how the corralling of a government has taken on a new phase that involves the corralling of climate science. The emergence of the strategy of corralling climate science may be a unique historical development that is associated with the emergence of a risk society and broad awareness of the environmental implications of industrial society (Beck, 1999). The global threat identified by scientists requires a series of interlocking technological transitions from electricity, buildings, and transportation based on fossil fuels to ones based on low-carbon energy alternatives. However, the dynamic of reflexive modernization engenders counter-movements of anti-reflexivity, which are funded by industries that face sunsetting due to the transition (Dunlap & McCright, 2011).

In the United States, the transition to a low-carbon energy infrastructure is a long-term, variegated political process that has been occurring since the 1970s, and there are various points of intersection with the trend toward market-oriented regulation that has also occurred during the period. During the 1970s, government-supported energy research and industrial development diversified from nuclear and fossil fuels to include renewable energy and energy conservation in response to the growing environmental movement, broad public concerns with air quality, and the oil crisis of 1973 (Hess, 2012). Although President Reagan slashed the budgets for renewable energy and energy conservation, they were never eliminated after the 1970s. The 1978 Public Utilities Regulatory Policies Act represented an early step toward a market-oriented restructuring of the electricity industry, because the law opened up electricity generation to small generators by requiring utilities to purchase electricity from them if their price was below the avoided cost (Hirsh, 1999). More comprehensive liberalization of electricity markets occurred during the 1990s, when electricity restructuring enabled competition among both electricity generation companies and retail distributors (*ibid.*).

During the 1990s, state governments also began to develop policies that encouraged the transition of electricity generation (Hess, 2012). In the process, the renewable energy industry shifted from a small niche oriented toward home power enthusiasts and off-grid generation to utility-scale solar and wind farms and to the sometimes larger niche of distributed (roof-top) generation. The technological system underwent a process of incorporation into grid-based energy production, and the technologies also underwent design transformations to accommodate their integration (on the "incorporation and transformation" process more generally, see Hess, 2007). Various market-oriented policies emerged from the restructuring of the electricity industry: competitive grants for renewable energy projects, capital assistance for start-up companies, renewable portfolio standards that created markets for renewable electricity, and regional carbon trading programs.

The corralling of climate science crystallized after another set of reform efforts during the late 1990s. In 1997 the Clinton administration agreed to join the Kyoto Protocol, which would have required the United States to reduce its greenhouse gas emissions to approximately 5% below 1990 levels by the year 2012. However, under the Byrd-Hagel Resolution of 1997, the Senate failed to ratify the treaty agreement. The oil industry had worked to stop climate-related policy reforms since at least 1989, when Exxon-Mobil and the American Petroleum Institute formed the Global Climate Coalition, which promoted climate science denialism (Rahm, 2010). After the Kyoto Protocol, several large companies accepted the scientific consensus and withdrew from the Global Climate Coalition, but Exxon-Mobil then formed the Global Climate Science Team to promote denialism (*ibid.*). When George W. Bush became president in 2000, he confirmed that the United States would not join the Kyoto Protocol and stated that there was not enough evidence in support of climate mitigation policy.

Although the Energy Policy Act of 2005 and the Energy Security and Independence Act of 2007 included some support for renewable energy and energy efficiency, the administration of President George W. Bush favored fossil-fuel development. Consequently, policies in support of the green-energy transition continued to be developed largely at the state-government level. During the 2000–2008 period of the Bush presidency, states with Democratic Party majorities developed multiple policies that facilitated both the development of green innovation clusters and demand for energy efficiency and renewable energy (Hess, 2012). Republican governors such as Arnold Schwarzenegger, Tim Palwenty, George Pataki, and even Rick Perry also facilitated the growth of renewable energy, albeit at different levels of enthusiasm. Bipartisan initiatives among governors supported regional climate change policy development in the Northeast, Midwest, and West. Likewise, during the 2008 presidential election both Republican and Democratic presidential candidates supported federal climate change legislation.

The political situation changed with the election of President Obama, who embraced a transition in the energy basis of the economy as a central part of his presidency. Obama's promise to create five million "green jobs" enabled him to build support among the party's base of progressives, environmentalists, the urban poor, and labor unions and to reach out to other voters who were concerned with employment and job security. Once elected, climate science and green-energy research went from a highly marginal position, in which government scientists faced censorship, to the centerpiece of a green industrial policy initiative. Secretary of Energy Steven Chu embraced the science of climate change and the need for a technological transition to renewable energy and energy efficiency, and the Department of Energy became the site for the administration's new green industrial policy (Hess, 2012). The American Recovery and Reinvestment Act provided funding for green-energy business development, consumer demand for renewable energy and energy efficiency, and training programs for green jobs. The House of Representative's American Clean Energy and Security Act (HR 2545) addressed the demand side of green industrial policy by supporting carbon regulation and a national renewable energy and energy efficiency standard (20% by 2020). In the Senate, John Kerry and Joseph Lieberman supported their version of the House cap-and-trade bill, the Clean Energy Jobs and America Power Act (S. 1733), with a diverse coalition that included retired military officers, leaders of some large corporations, the renewable energy industry, unions, and some environmentalists.

However, by the summer of 2010 it was clear that demand-side policy (carbon regulation and a national renewable electricity standard) of green industrial policy would not survive a vote in the Senate. The Kerry-Lieberman bill was defeated due largely to a coalition of political conservatives and the fossil-fuel industry. The lobbying strength of the oil and gas industry ($146 million in 2012), the utility industry ($141 million), and mining ($32 million, much of it from the coal industry) far outpaced the capacity of the countervailing renewable energy organizations, such as the American Wind Energy Association ($2.4 million) and the Solar Energies Industry Association ($1 million; Center for Responsive Politics, 2013). The billionaires David and Charles Koch developed a network of conservative and fossil-fuel donors who helped to provide support for the anti-green Republican backlash that became apparent in the midterm election results of November 2010 (Eilperin, 2012, Fang, 2010). They also joined with other fossil-fuel donors to support and publicize climate change denialism and skepticism (Greenpeace, 2010).

The transformation of the policy field was evident in the 2010 elections. Ninety-four of the 100-person entering class of Congress explicitly denied the science of climate change, signed the "No Climate Tax Pledge" of Americans for Prosperity (a group funded by the fossil-fuel billionaires Charles and David Koch), and/or signed the "FreedomWorks Tea Party Contract with America," which opposes cap-and-trade legislation (Johnson, 2010). About half of all Republican members of both houses had made statements indicating that they questioned the science or scientific consensus about climate change (*ibid.*).

Many of the Republican governors and state legislators who were newly elected in 2010 made dramatic reversals on climate-related policy in cases where they replaced Democratic governors. For example, Governor Chris Christie of New Jersey withdrew the state from the Regional Greenhouse Gas Initiative, and Maine's governor proposed a long series of environmental roll-back initiatives that led to a battle with the state's legislature. In Wisconsin, Governor Scott Walker worked to roll back the state's wind energy development program, and in Colorado, Michigan, Montana, Ohio, Washington, and West Virginia legislators attempted to end the state's renewable portfolio standard laws, which dedicated a portion of the state's electricity from renewable energy sources (Hess, 2012).

However, in states with a strong clean-tech industry, such as Iowa and Nevada, Republican governors adopted more moderate positions. Likewise, in states with Democrats as governors, new legislation in support of green energy continued to win legislative approval. In California a coalition of unions, environmentalists, civil rights, renewable energy industry, and green venture capital leaders provided the organizational strength and financial support to turn back a ballot assault on the state's AB 32 law, which had set in motion the mechanism for carbon regulation (Hess, 2012). Likewise, in that year then Attorney General Jerry Brown won the governor's office, and he openly support green-energy transition policies within the context of limited budgets.

CLIMATE DENIALISM AND PARTY CONFLICT

One of the central political actors that supported climate denialism was the Tea Party movement. Climate change denialism and skepticism has both a longer history than the existence of the Tea Party movement and a broader range of political support, but the Tea Party movement became a major vector that brought climate change denialism and skepticism into state legislatures, governors' offices, and the US Congress. Although the central frame of the movement was government overspending, Tea Party candidates who came into office in 2010 also had strong anti-environmental stands that were consistent with the network of fossil-fuel and conservative donors that supported their candidacies. For example, a survey of self-identified Tea Party adherents showed that they generally did not believe in global warming and did not support a 20% renewable energy standard or a global carbon trading agreement, whereas the majority of Democrats, Republicans, and independents held the opposite views (Leiserowitz, Maibach, Roser-Renouf, & Hmielowski, 2011). The backlash continued in 2011 and 2012 during the Republican Party nomination for the presidency, when candidates who had formerly supported green transition policies in their states reversed their positions in order to win the nomination.

Although there was a close relationship between climate change denialism and conservative political candidates, there was also a general a decision among Republican Party leaders in Congress to undercut President Obama's policies by denying him legislative victories and by branding him as a big-spending social liberal or, in the case of health-care reform, a socialist. Even moderate Republicans who accepted the scientific consensus on global warming faced primary battles and attacks from the anti-environmental right wing of their own party. In general, relatively moderate Republicans who had formerly supported climate change policy, such as Senator John McCain, quieted their views and came in line with the party position, and Republican Party support for climate change became increasingly limited to retired leaders. Former Senator John Warner, former Representative Sherwood Boehlert, former Representative John Inglis, and former Secretary of State George Schultz are examples of Republican retirees from government leadership positions who defended the need for climate change action (Davenport, 2011).

By 2011, Republicans had targeted the failed investments in solar energy companies such as Solyndra to launch a general attack on the green developmentalism of the Obama administration and to defend a neoliberal alternative that left investment decisions to private markets (Hess, 2012). In response to the attacks on his green transition policies, the president backed away from the strong green jobs rhetoric that had characterized his 2008 election. His initial presidential campaign speeches in late 2011 focused on jobs but carefully deleted references to green jobs, and a study of his speeches showed that in general he had backed away from green-energy rhetoric (Roberts & Kincaid, 2012). Nevertheless, he moved ahead with green-energy policies in non-legislative arenas, such as by supporting changes in fuel efficiency in the automotive industry and by facilitating the transition to renewable energy in the military (Hess, 2012). He also backed increases in green-energy research and development in his budget proposals, and in January, 2011, the Environmental Protection Agency also began to require that big polluters would have to obtain permits for greenhouse gas emissions. The rules were focused on electricity generation plants and oil refineries, but they were scheduled to be extended to other types of industrial polluters.

Republicans in the House responded by grilling the EPA administrator Lisa Jackson, by arguing that carbon regulation went beyond the agency's legislative mandate, and by proposing budget cuts that would reduce the agency by one-third. Some Republican presidential candidates even promised to abolish the agency. The fossil-fuel industry also launched lawsuits against the EPA, and in concert with conservative think tanks the industry also developed plans for a coordinated national attack on wind energy (Goldenberg, 2012). The plan included attempts to roll back state-government legislation, support for NIMBY struggles against wind farms, and a $6 million purchase of advertising against President Obama's green-energy plans. In the various attacks on green-energy policy, neoliberal ideology was often a point of reference and a way of distinguishing Republican Party conservativism from the Obama administration's developmentalism.

MECHANISMS OF CORRALLED SCIENCE

As opposition to green-energy reforms grew, the efforts to corral climate science also increased. The corralling of climate science in the United States during this period involved three main mechanisms, the first of which is intellectual suppression. Widely recognized in the sociology of science literature, intellectual suppression is especially common for individual scientists who raise questions about public health or environmental risks associated with exposure to industrial chemicals and other products (e.g., Delborne, 2008; Martin, 2010; Martin, Baker, Manwell, & Pugh, 1986). The most visible scientist leaders of the climate science counterpublic have suffered from personal attacks on Internet web sites and in the media, attacks on their research by contrarian scientists, and intimidation such as an envelope with suspicious white power (Clynes, 2012). The scientists have also been subjected to hostile lawsuits, Freedom of Information Act requests, and investigations by Congress and universities (*ibid.*). Although some climate scientists have chosen to provide public rebuttals to climate science deniers and skeptics, knowledge about the intimidation has encouraged others to stay out of the public spotlight. Thus, the prospect of exposing oneself to intellectual suppression motivates some climate scientists to maintain a low profile, stay out of the public limelight, and avoid becoming involving with politics and policy.

However, the strategy of suppression can create a secondary dynamic of backfire (Martin, 2007). In other words, attacks on scientists can generate media coverage (e.g., Clynes, 2012), and the public limelight can turn climate scientists under attack into public figures. They can become portrayed as heroic figures who are defending the light of knowledge and public interest against bullying figures who represent special interests. When the media coverage increases, scientists are given a platform to showcase their rebuttals to the unfounded arguments of critics. Once they are in the public eye, they may become more willing to expand their role from that of making scientific statements (what is the case) to that of giving opinions on policy options, such as James Hansen has done in his support for the carbon tax (Hansen, 2009). Furthermore, the suppression of individual scientists has helped to motivate the broader climate science community to respond to attacks (e.g., Trenberth, 2012). Their messages imply the need for policy action and the weakness of a neoliberal approach to the problem, which would leave the green transition to long-term market forces.

The second mechanism of the corralling of science involves journalists and policymakers who produce disinformation about the state of scientific controversy among experts. In the case of climate science denialism, there were two phases to this mechanism. Climate change denialists in right-wing think tanks have been funded for some time, and they have produced a perception in the media of a higher level of controversy among scientific experts than actually took place (Dunlap & McCright, 2011; Fang, 2010; Oreskes & Conway, 2010). Furthermore, the elimination of moderate Republican candidates in primary elections in 2010 and 2012, and the general shift in the Republican Party's energy policy toward a pro-fossil-fuel stance, magnified the public dissemination of the denialist message by providing spokespersons in the political field.

Again, this mechanism produces a countervailing dynamic that tends to undermine climate denialism in the media and political field. There is growing recognition that the media standard of achieving a "balanced" story has led to the magnification of the perspective of contrarian scientists who may have some general credentials but lack standing in the climate science field (Boykoff, 2013). In turn, the recognition has led some journalists and advocates to call into question the air space given to contrarian scientists as itself a form of misinformation. For example, the television network CNBC has been criticized for its willingness to air climate denialist perspectives at a rate of over 50% of stories during 2012–2013 (Strupp, 2013). The fact-checking web site Media Matters has highlighted how the continued misuse of the tradition of the balanced story has led to divisions among journalists, such as a panel of senior business journalists who criticized the coverage of CNBC (*ibid.*).

A second countervailing dynamic also results from the success of climate denialism in the political field. Although it has been possible to purge Republican moderates in primary campaigns, the extreme position of Tea Party and other highly conservative candidates on climate and environmental issues can put them at a disadvantage in some general elections. The election of a climate-supporting governor over a climate-denying candidate in Virginia in 2012 has sometimes been cited as an example of how climate denialism does not play well across a wide range of voters in a general election. In support of this proposition, Krosnick, MacInnis, and Vilar (2011) compared voting preferences for a hypothetical senate candidate who held one of the following three positions: pro-green, not green (i.e., denies climate science and the need to respond to it), or no position on climate change. For Democrats, the preferences were 74% in favor of the pro-green candidate, 37% in favor of the not green candidate, and 53% in favor of the candidate with no position. Independents had a pattern similar to that of Democrats. For Republicans, the differences in preferences were not statistically significant and all in the 76–83% range. For all affiliations together, the preferences were 77% pro-green, 48% anti-green, and 65% no position. The results suggest that taking a pro-green position will help both Democrat and Republican candidates in general elections. Consistent with those data, a poll conducted for the League of Conservation Voters in 2013 showed that voters for senate candidates in swing states prefer candidates who support environmental and clean-energy issues (Hart Research Associates, 2013).

The third mechanism of the corralling of science is the systematic reduction of government funding for agencies that sponsor the countervailing research. On October 14, 2011, Ralph Hal – the chair of House Science, Space, and Technology – and Republican colleagues issued a statement of intent to cut funding for programs related to climate science and green energy. The cuts were proposed to take place across agencies and represented a broad attempt to end the federal government's support of climate change research (Morello, Maron, Freidman, & Rahim, 2011; US House of Representatives, 2011). The cuts included the following:

- Reduce by one third the Department of Energy's Energy Efficiency and Renewable Energy portfolio.
- End ARPA-E funding because of its "venture capital" investments and consider funding only for very early stage projects.
- Cut the Atmospheric System Research and the Climate and Modeling Program of the Office of Research of the Department of Energy.
- Shift the DOE's fossil-fuel research from carbon sequestration to oil and gas exploration.
- Cancel NASA's Orbiting Carbon Observatory satellite.
- Reduce by 20% the program within NASA's Earth System's Mission account, which includes research related to climate change.
- Cut the National Science Foundation contributions to the Climate Change Technology Program and US Global Research Program, and eliminate the Climate Education Program.
- Cut the funding for the Air, Climate, and Energy programs and for the Integrated Risk Information System Program of the Environmental Protection Agency.

Although many of the proposed cuts were not approved by the Senate, which was controlled by Democrats, some proposals were successful. For example, NASA's climate-related science budget was cut, and it was forced to cancel grants (Clynes, 2012; Vastag, 2011). In 2012, the House of Representatives also voted to end the budget for the Political Science Program of the National Science Foundation. Among the grants mentioned was an agent-based model of climate change (Jones, 2012).

We do not yet know the general effects on research proposals and funding patterns, but there are some cases of undone science that have already been produced by the cuts. For example, the National Oceanic and Atmospheric Administration froze the "Twentieth Century Reanalysis" project, which provides hour-by-hour changes in the atmosphere during the twentieth century (Morello, 2012). Furthermore, scientists and programs across the government were attempting to avoid mentioning climate change in their proposals (Clynes, 2012).

For this mechanism there are two primary countervailing dynamics. Within the government, attempts to cut climate science funding produce mobilizations by Congressional Democrats, who can restore the cuts in final budget negotiations, and by the Democratic president (Semeniuk,

Tollefson, Wadman, Mann, & Dickey Zakaib, 2011). The affected executive branch agencies are also generally unhappy with the interference, and they may devise work-arounds to the cuts. But the cuts also produce a secondary dynamic outside Congress because scientists and universities will mobilize to preserve research funding. For example, professional associations in the social sciences mobilized to resist the attack on political science and to head off future attacks on other social sciences, and climate scientists lobbied Congress in their annual Climate Science Day event. Likewise, university presidents have responded to threats posed by the general budget by lobbying Congress for continued funding especially in light of the budget sequester that occurred in 2013 (Anderson, 2013). The ongoing mobilization by scientific associations and universities is not likely to have the same lobbying capacity of the fossil-fuel industry, but it only needs to be strong enough to block further cuts by recruiting sufficient allies in Congress and the executive branch.

In summary, epistemic rift – the breakdown of the exchange of government support for scientific expertise – occurs through three mechanisms: the suppression of scientists, the propagation of denialism in the media and among elected officials, and the cutting and sequestering of government funding. A segment of the political leadership that is significant enough to have policy effects rejects the consensus knowledge of the scientific research community and targets scientists and research programs that are aligned with the rejected knowledge. These mechanisms of epistemic rift in turn generate countervailing dynamics, such as backfire for the suppressed scientists, credibility gaps for journalists and elected officials who broadcast the views of contrarian scientists, and mobilizations by professional associations and universities to restore funding. The counter-dynamics help to equilibrate the levels of funding invested by the coalitions that support contrarian science and its propagation through think tanks, the media, and political advertising, because the counter-dynamics generate unpaid media attention that can showcase the misleading information that is circulating. Although the system could reach an equilibrium in which all government-supported climate science is defunded and the mainstream media coverage of contrarian science approaches 100%, this outcome seems unlikely. Although the Catholic Church could refuse to look through the new telescopes and could opt to suppress Galileo, in the long run it was not able to stop the emerging astronomical knowledge of the scientific field. In a similar way, the more likely outcome of the interaction of the corralling mechanisms and the countervailing dynamics is the re-education of the public through the media coverage of information manipulation and the eventual collapse of the situation of epistemic rift. However, in the case of climate science this outcome may take years, and each passing year of delay generates new long-term risks and the prospects of more expensive and painful adaptation.

CRACKS IN THE CORRAL

Is there evidence that the corralling of climate science is losing its effectiveness? Public opinion polls indicate that the effects of climate denialism and skepticism on public opinion peaked in 2010 and began to reverse. For example, Republicans who said that they believed there was scientific evidence of global warming declined from 59% in 2006 to 35% in 2009 but then increased in 2011 to 43%, and there was a similar bottoming out and uptick for Independents. Among moderate and liberal Republicans (who represent about a third of all Republicans), those who stated that there was solid evidence for global warming grew from 41% in 2009 to 63% in 2011, whereas among conservative Republicans opinion remained constant at 31% during the period (Pew Research Center, 2011). In the survey of persons of all political persuasions, belief in the reality of global warming continued to grow between 2011 and 2013, from approximately 50% to 69%, and belief in the anthropogenic cause of global warming also grew, albeit at a lower level that reached 42% in 2013 (DeSilver, 2013).

Why did public opinion on climate denialism and skepticism bottom out and start to recover its previously higher levels? Specific events have some effect on public opinion. For example, the "Climategate" stories in November, 2009, which were released to undermine the credibility of climate scientists at the time of United Nations Copenhagen summit, had some effect on public opinion, although the effect was limited by awareness of the story. In a survey conducted in December, 2009, and January, 2010, Leiserowitz, Maibach, Roser-Renouf, Smith, and Dawson (2010) found that only 29% of respondents had heard of the Climategate news story, and in a survey in June, 2010, Krosnick (2010) found that only 9% of Americans knew about it. However, Leiserowitz and colleagues also found that among the respondents who had followed the story, about 47% said the stories made them somewhat or more certain that global warming was not happening. Krosnick and colleagues also showed that including a skeptic in a news story negatively affected belief in global warming (Malka, Krosnick, Debell, Pasek, & Schneider, 2009).

If media portrayal of climate denialism and skepticism and stories sympathetic to that viewpoint negatively affects public opinion, one would expect that similar stories in the other direction would positively affect public opinion and might explain the uptick in public belief after 2009. For example, one explanation of the uptick in public belief may be a reaction to weather. High outdoor temperatures have been shown to increase belief in global warming (Joireman, Truelove, & Buell, 2010), and there were many stories after 2009 about record-breaking heat waves and other unusual weather events. Another explanation is the possible effect of media stories favorable to climate science that appeared in 2011 and 2012, but there are no studies at this point that confirm the effects of those stories on general public opinion. Four of the most significant events are as follows:

- In October, 2011, physicist Muller (2011) of the University of California at Berkeley released the results of a study that had been partly funded by the Charles G. Koch Foundation and was expected to be another media event in support of climate change denialism and skepticism. After reviewing 1.5 billion temperature readings, the team concluded that global temperature had increased by about 0.7°C since 1957, a finding that confirmed consensus science on climate change.
- In February, 2012, climate scientists released a widely cited reply to an op-ed editorial in the *Wall Street Journal* in which they argued that the so-called climate scientists who signed a previous denialist editorial were mostly outside their area of expertise, a phenomenon that they likened to "dentists practicing cardiology" (Trenberth, 2012).
- In February, 2012, Peter Gleick, a scientist who had been awarded a MacArthur Fellowship for work on water issues, used an assumed name and obtained documents from the conservative Heartland Institute, which had sponsored conferences of global warming skeptics and deniers. The documents revealed information about the organization's donor connections, financial support for climate change deniers, and other activities. Although the organization claimed that at least one of the documents was forged, and the event backfired on Gleick, it did provide a great deal of media coverage of the funding of denialism.
- In May, 2012, the same organization became embroiled in another media controversy over an advertising campaign that likened supporters of climate science to terrorists. The backfire from the campaign led many corporate donors to withdraw support, including State Farm and General Motors. At the annual conference of climate denialism, the institute announced that it lacked funds to hold another annual conference (Lacey, 2012).

In 2012, the re-election of President Obama and the continued control of the US Senate by Democrats, as well as extreme weather events such as the Midwestern drought and coastal storm Sandy, may have taken some of the winds out of the sails of climate denialism. It is possible that the Republican Party and fossil-fuel industry will conclude that that support for climate change denialism and skepticism is a waning phenomenon and politically damaging in elections in districts outside deeply conservative regions. Climate change denialism may become one part of a group of conservative, "white male" issues that are losing political ground in the country due to broader demographic changes. The 2012 Republican candidate Mitt Romney may have served as a bellwether of the shift. He stated on several occasions that he believed global warming was real and that it was at least partially caused by humans. Although there were some occasions when he waivered or even expressed denialism and skepticism, in general his position was to sidestep the science issues and ground an anti-green-energy position on economic arguments. On his presidential campaign web site, the page on "energy" voiced strong opposition to Obama's "green jobs" programs and supported continued development of fossil fuels (Romney, 2012); however, the web site also showed support for ARPA-E as an example of bipartisan, basic research, a view that was at odds with that of the House Republicans discussed above. In his campaign speeches, Romney also opposed cap-and-trade legislation unless other countries, such as China, also agreed to it (Jacobson, 2012). Thus, although he remained within the party mainstream on the policy issues, he was not a vociferous exponent of climate denialism.

It is possible that the epistemic rift that occurred in the political field will be undermined by the failure to maintain a similar epistemic rift in general public opinion, and as a result political candidates in closely contested races will moderate their views in order to appeal to a wide range of voters. If the trend continues in favor of the regrowth of public confidence in climate science, and if denialist candidates find that their political future is harmed in general elections, then we might see the collapse of climate denialism as a political force. In such a situation, we might see the closing of the epistemic rift and the re-emergence of the doxological function of the climate science research field.

However, this change would not necessarily mean that the political opportunity structure for green transition policies would automatically reopen. A possible harbinger of the next phase was a speech in June, 2012, by the CEO of Exxon-Mobil, Rex Tillerson, who stated that global warming was real and that fossil fuels were contributing to it, but he added that fears were overblown and society that would adapt (Tillerson, 2012).

Tillerson's use of the word "adapt" points to another development in the policy field and a possible reconfiguration of the epistemic rift. Across the world and across the United States, the policy responses to climate change have diversified from mitigation strategies to adaptation strategies. According to a study by the PEW Center on Global Climate Change (Cruce, 2009), by 2009 18 states had launched or begun adaptation plans, and our review in 2013 indicated that more plans had emerged since then. Likewise, the US federal government and associated organizations have been developing climate adaptation plans. For example, the US Department of Transportation (2012) published studies on the potential effects on climate change on transportation infrastructure, from buckled rails to flooding, and the National Academies of Sciences published a review of adaptation issues for airports (Airport Cooperative Research Program, 2012). Most of the policy initiatives in this area address concerns with water, such as droughts, flooding, and sea water rise. In effect, water is to climate adaptation as energy is to climate mitigation.

Water concerns have a different epistemic weight than do those for energy. In the case of climate mitigation, energy technologies must be transitioned to low-carbon forms, and a powerful industrial sector is threatened by declining opportunities. In the case of water, the relationship to climate change is much less direct, because water-related resilience is complicated by a variety of human-caused factors, such as decisions regarding flood plain development, agricultural land use, deforestation, suburban sprawl, general population growth, groundwater use, and the capacity of infrastructure to absorb rain water. As a result, political leaders can support adaptation policies without necessarily taking a stand on the relative importance of climate change in posing general risk. Furthermore, the technical difficulties of downcasting climate models to an urban or even state-government scale also make climate scientists hesitant to apply their research at a local level except in the most general terms. Finally, if adaptation policies are not combined with climate mitigation, they can end up supporting the continued use of fossil fuels under the presumption that societies will adapt to the changes.

Thus, adaptation can serve as a boundary object across various constituencies. For the environmentalists, it can be an unfortunate outcome of failed mitigation policies, an opportunity to design systems that are both adaptive and sustainable, and an occasion to reveal the importance of a double strategy of mitigation and adaptation. For the opponents of mitigation strategies, adaptation can simply mean, "Live with it," as the ExxonMobil CEO was suggesting. Although it is far from certain what role adaptation will play as part of the continued strategy of the fossil-fuel industry to slow the green-energy transition, it represents one possible outcome in which there is a return to something approximating the standard model of science-policy relations. In other words, the fact of global warming can become widely accepted as the doxa of the political field, but the policy implications – the relative balance of mitigation versus adaptation investments – would remain highly contested politically. This outcome of the interaction of corralling mechanisms and countervailing dynamics appears likely even if it may not be ideal.

CONCLUSION

The epistemic rift for climate science in the United States is not only of grave political and environmental importance, but it is also an occasion to develop theory in the political sociology of science and technology. In previous research, I focused on the alternative counterpublic relationship, in which scientists are marginalized institutionally and/or epistemically because of their advocacy of research programs that are outside the mainstream of their research fields. The paradigm case here was the networks of scientists, clinicians, patients, and nutraceutical companies that advocated the development of non-toxic, nutrition-based therapies for chronic disease. In the full-blown scientific counterpublic, scientists also have a relationship with actors in the subordinate positions in the political and industrial fields, such as grassroots social movement organizations and entrepreneurs who develop alternative technologies (Hess, 2011). Together, scientists and activists draw attention to the pattern of undone science, or the systematic underfunding of research that would address the knowledge questions of the counterpublic and the need for alternative technologies and products. Thus, they question the agendas of the dominant research fields and the consensus among regulators, industry, and mainstream scientists. This type of relationship is common in the complementary and alternative medicine fields as well as for many environmental health issues in which a counterpublic draws attention to risk and uncertainty, often in opposition to assurances of safety from industry.

In the case of climate science, the scientists who attack the consensus on the anthropogenic forcing of climate change are themselves in marginal positions at least with respect to the specific field of climate science. Some of the climate science deniers lack credentials as natural scientists, and even the natural scientists generally lack credentials as climate scientists and are not active producers of research; hence, they lack standing within the relevant research field. Some of them have also been documented to have ties to industry-funded contrarian science campaigns (e.g., Oreskes & Conway, 2010). Although there are controversies in the climate science research field, the controversies occur within a shared set of assumptions that contemporary global warming is established and that anthropogenic greenhouse gases play a very significant role in global warming. Thus, a scientist like James Hansen is within the mainstream of the research field, even if he suffers suppression in the political field and arrests at demonstrations (Hansen, 2009). Conversely, whereas scientists who express climate skepticism and denialism lack symbolic, temporal, and other forms of capital in the scientific field, they have standing with powerful corporations, which in turn work to restructure the scientific field (by suppressing scientists and cutting funding) and the political field (by purging the Republican Party of leaders who are soft on climate change and by pouring money into think tanks and media advertising).

The study of the politics of climate science in the United States suggests that countervailing dynamics will emerge from the attempt to corral a research field that has achieved a high level of consensus and has made important recommendations to policymakers. As the public becomes aware that so-called climate "skeptics" are not credentialed climate scientists and that most of the scientists in the research field share a consensus about the role of anthropogenic forcing of global warming, then the denial machine begins to lose its effectiveness at corralling the science, and the epistemic rift may begin to close up. It is possible that the sharp rise in climate denialism witnessed after the 2010 elections began to decline by 2012, when the controversies over the Heartland Institute garnered media attention, the admission by the CEO of Exxon-Mobil that global warming is real undermined denialism, and extreme weather events led to greater awareness of the possibility of climate change.

As the negative publicity on the harassment of climate scientists increases, weather records continue to be broken, and public opinion in support of the reality of global warming and greenhouse gas causation increases, then Republican Party leaders may shift the strategy from denialism and skepticism about the science to a different kind of pro-fossil-fuel politics. This situation would be a return to the standard model for the science-policy relationship, in the sense that the consensus knowledge of the research community would form the epistemic ground on which political debate would take place. However, as argued above, the return of the doxological function of climate science in the political field would not necessarily imply that policy action on climate mitigation would be imminent. Instead, the debate could turn more on what, if anything, should be done about the "fact" that has now recovered its credibility in the political field. As the CEO of Exxon-Mobil suggested, the response from the fossil-fuel industry is to urge adaptation rather than mitigation. Indeed, adaptation appears to be the dominant direction of policy reform for the twenty-first century. Unfortunately, the prospect of adapting to climate change is likely to be especially brutal for the world's most vulnerable human populations, not to mention other species.

NOTE

1. The terms "denialism" and "skepticism" are used together to indicate a spectrum of opinion from self-conscious denial of established scientific research to questions about some aspects of climate science that are based on lack of knowledge.

REFERENCES

Airport Cooperative Research Program. (2012). *ACRP synthesis 33: Airport climate adaptation and resilience*. Washington, DC: Transportation Research Board of the National Academies of Science.

Anderson, N. (2013). Universities continue to lobby against sequester's cuts of research funding. *Washington Post*, November 13, n.p. Retrieved from http://www.washingtonpost.com/local/education/universities-continue-to-lobby-against-sequesters-cuts-of-research-funding/2013/11/12/64c29e68-4bba-11e3-be6b-d3d28122e6d4_story.html. Accessed on December 30, 2013.

Barley, S. (2010). Building an institutional field to corral a government: A case to set an agenda for organization studies. *Organizational Studies*, *31*(6), 777–805.

Beck, U. (1999). *World risk society*. Malden, MA: Polity Press.

Bourdieu, P. (2001). *Science of science and reflexivity*. Chicago, IL: University of Chicago Press.

Bourdieu, P. (2005). *The social structures of the economy*. Malden, MA: Polity Press.

Boykoff, M. (2013). Public enemy number 1: Understanding media representations of outlier views on climate change. *American Behavioral Scientist*, *57*(6), 796–817.

Brown, P. (2007). *Toxic exposures: Contested illnesses and the environmental health movement*. New York, NY: Columbia University Press.

Camic, C. (2011). Bourdieu's cleft sociology of science. *Minerva*, *49*(3), 275–293.

Center for Responsive Politics. (2013). Ranked sectors. Retrieved from http://www.opensecrets.org/lobby/top.php?indexType=c. Accessed on December 30, 2013.

Clynes, T. (2012). The battle. *Popular Science*, July, pp. 36–43, 80.

Cruce, T. (2009). *Adaptation planning: What US states and localities are doing*. Pew Center on Global Climate Change. Retrieved from http://www.c2es.org/docUploads/state-adapation-planning-august-2009.pdf. Accessed on December. 30, 2013.

Davenport, C. (2011). Retired republicans quietly try to shift GOP climate-change focus. *National Journal*, October 7. Retrieved from http://www.nationaljournal.com/energy/retired-republicans-quietly-try-to-shift-gop-climate-change-focus-20110930. Accessed on December 30, 2013.

Delborne, J. (2008). Transgenes and transgressions: Scientific dissent as heterogeneous practice. *Social Studies of Science*, *38*(4), 509–541.

DeSilver, D. (2013). *Most Americans say global warming is real, but opinions split on why*. Pew research Center, June 6. Retrieved from http://www.pewresearch.org/fact-tank/2013/06/06/most-americans-say-global-warming-is-real-but-opinions-split-on-why/. Accessed on December 30, 2013.

Dunlap, R., & McCright, A. (2011). Climate change denial: Sources, actors, and strategies. In C. Lever-Tracy (Ed.), *Routledge handbook of climate change and society*. New York, NY: Routledge.

Eilperin, J. (2012). Climate skeptics target energy law. *Washington Post*, November 25, p. A4.

Fang, L. (2010, October 20). *Memo: Health insurance, banking, oil industries met with Koch, Chamber, Glenn Beck to plot 2010 election*. Retrieved from http://thinkprogress.org/politics/2010/10/20/124642/beck-koch-chamber-meeting/. Accessed on December 30, 2013.

Fligstein, N., & McAdam, D. (2012). *A theory of fields*. Oxford: Oxford University Press.

Foster, J. B. (2000). *Marx's ecology: Materialism and nature*. New York, NY: Monthly Review Press.

Foucault, M. (2007). *Security, territory, population: Lectures at the college de France, 1977–1978*. London: Palgrave Macmillan.

Foucault, M. (2008). *The birth of biopolitics: Lectures at the collège de France, 1978–1979*. London: Palgrave Macmillan.

Frickel, S., Gibbon, S., Howard, J., Kempner, J., Ottinger, G., & Hess, D. (2010). Undone science: Social movement challenges to dominant scientific practice. *Science, Technology, and Human Values*, *35*(4), 444–473.

Frickel, S., & Gross, N. (2005). A general theory of scientific/intellectual movements. *American Sociological Review*, *70*(2), 204–232.

Goldenberg, S. (2012). Conservative think tanks step up attacks against Obama's clean Energy strategy. *The Guardian*, February 7, n.p. Retrieved from http://m.guardian.co.uk/environment/2012/may/08/conservative-thinktanks-obama-energy-plans?cat=environment&type=article. Accessed on December 30, 2013.

Greenpeace. (2010). *Koch industries secretly funding the climate denial machine*. Retrieved from http://www.greenpeace.org/usa/Global/usa/report/2010/3/executive-summary-koch-indus.pdf. Accessed on December 30, 2013.

Greenwood, J. (2002). *The effectiveness of EU business associations*. New York, NY: Palgrave.

Guston, D. (2001). Boundary organizations in environmental policy and science: An introduction. *Science, Technology & Human Values*, *26*(4), 399–408.

Hansen, J. (2009). *Storms of my grandchildren: The truth about the coming climate catastrophe and our last chance to save humanity*. New York, NY: Bloomsbury USA.

Hart Research Associates. (2013). *Voters attitudes about EPA carbon regulations and climate change*. Retrieved from http://www.lcv.org/issues/polling/11-state-polling-memo.pdf. Accessed on December 30, 2013.

Hess, D. (2007). *Alternative pathways in science and technology: Activism, innovation, and the environment in an era of globalization*. Cambridge, MA: MIT Press.

Hess, D. (2009). The potentials and limitations of civil society research: Getting undone science done. *Sociological Inquiry*, *79*(3), 306–327.

Hess, D. (2011). To tell the truth: On scientific counterpublics. *Public Understanding of Science*, *20*(5), 627–641.

Hess, D. (2012). *Good green jobs in a global economy*. Cambridge, MA: MIT Press.

Hess, D. (2014). Beyond scientific consensus: Scientific counterpublics, countervailing industries, and competing research agendas. In W. Viehover & P. Wehling (Eds.), *The public shaping of medical research: Patient associations, health movements, and biomedicine*. London: Routledge.

Hicks, A., & Lechner, F. (2005). Neopluralism and neofunctionalism in political sociology. In T. Janoski, R. Alford, A. Hicks, & M. Schwartz (Eds.), *The handbook of political sociology* (pp. 54–71). Cambridge: Cambridge University Press.

Hirsh, R. (1999). *Power loss: The origins of deregulation and restructuring in the American electric utility system*. Cambridge, MA: MIT Press.

Irwin, A. (2006). The politics of talk: Coming to terms with the "new" scientific governance. *Social Studies of Science*, *36*(2), 299–320.

Jacobson, L. (2012). On mitt Romney and whether humans are causing climate change. *Tampa Bay Times*, October 11. Retrieved from http://www.politifact.com/truth-o-meter/statements/2012/may/15/mitt-romney/Mitt-romney-and-whether-humans-are-causing-climate/. Accessed on December 30, 2013.

Jacques, P., Dunlap, R., & Freeman, M. (2008). The organization of denial: Conservative think tanks and environmental skepticism. *Environmental Politics*, *17*(3), 349–385.

Jasanoff, S. (1994). *The fifth branch: Science advisors as policymakers*. Cambridge, MA: Harvard University Press.

Johnson, B. (2010). The climate zombie caucus of the 112th congress. *Think Progress*, November 19. Retrieved from http://thinkprogress.org/climate-zombie-caucus/. Accessed on December 30, 2013.

Joireman, J., Truelove, H., & Buell, B. (2010). Effect of outdoor temperature, heat primes, and anchoring on belief in global warming. *Journal of Environmental Psychology*, *30*(4), 358–367.

Jones, R. (2012). FY 2013 NSF political science research funding eliminated by house. *American institute of Physics Policy Bulletin 74*, May 25. Retrieved from http://www.aip.org/fyi/2012/074.html. Accessed on December 30, 2013.

Kinchy, A. (2012). *Seeds, science, and struggle: The global politics of transgenic crops*. Cambridge, MA: MIT Press.

Krosnick, J. (2010). *Large majority of Americans support government solutions to address global warming*. Stanford, CA: Woods Institute on the Environment, Stanford University. Retrieved from http://woods.stanford.edu/research/americans-support-govt-solutions-global-warming.html. Accessed on December 30, 2013.

Krosnick, J., MacInnis, B., & Vilar, A. (2011). *The impact of climate change on electoral success in 2010: Experimental evidence*. Stanford, CA: Woods Institute on the Environment, Stanford University. Retrieved from http://woods.stanford.edu/docs/surveys/Stanford_Climate_Politics2011.pdf. Accessed on December 30, 2013.

Lacey, S. (2012). The self-inflicted downfall of the heartland institute. *Think Progress*, May 24. Retrieved from http://thinkprogress.org/climate/2012/05/24/489430/the-self-inflicted-downfall-of-heartland-institute/. Accessed on December 30, 2013.

Lave, R. (2012). *Fields and streams: Stream restoration, neoliberalism, and the future of environmental science*. Athens, GA: University of Georgia Press.

Leiserowitz, A., Maibach, E., Roser-Renouf, C., & Hmielowski, J. (2011). Politics and global warming: democrats, republicans, independents, and the tea party. Yale Project for Climate Change Communications and George Mason Center for Climate Change Communication. Retrieved from http://environment.yale.edu/climate-communication/article/PoliticsGlobalWarming2011. Accessed on December 30, 2013.

Leiserowitz, A., Maibach, E., Roser-Renouf, C., Smith, N., & Dawson, E. (2010). Climategate, public opinion, and the loss of trust. Yale Project for Climate Change Communications and George Mason Center for Climate Change Communication. Retrieved from http://environment.yale.edu/climate/files/Climategate_Opinion_and_Loss_of_Trust_1.pdf. Accessed on December 30, 2013.

Lukes, S. (1974). *Power: A radical view*. New York, NY: Macmillan.

Mahoney, C. (2008). *Brussels versus the beltway: Advocacy in the United States and the European Union*. Washington, DC: Georgetown University Press.

Malka, A., Krosnick, J., Debell, M., Pasek, J., & Schneider, D. (2009). *Featuring skeptics in news media stories about global warming reduces public beliefs in the seriousness of global warming*. Stanford, CA: Woods Institute on the Environment, Stanford University. Retrieved from https://www.stanford.edu/dept/woods/docs/surveys/Global-Warming-Skeptics-Technical-Detail.pdf. Accessed on December 30, 2013.

Martin, B. (2007). *Justice ignited: The dynamics of backfire*. Lanham, MD: Rowman & Littlefield.

Martin, B. (2010). How to attack a scientific theory and get away with it (usually). *Science as Culture*, *19*(2), 215–239.

Martin, B., Ann Baker, C. M., Manwell, C., & Pugh, C. (Eds.). (1986). *Intellectual suppression*. London: Angus and Robertson.

McFarland, A. (2010). Interest group theory. In L. Maisel & J. Berry (Eds.), *The Oxford handbook of American political parties and interest groups* (pp. 37–56). Oxford: Oxford University Press.

Morello, L. (2012). NOAA halts reconstruction of past climate. *Scientific American*, February 10, n.p. Retrieved from http://www.scientificamerican.com/article.cfm?id=noaa-halts-reconstruction-past-climate. Accessed on December 30, 2013.

Morello, L., Maron, D. F., Freidman, L., & Rahim, S. (2011). Republicans gut EPA climate rules, slash deeply into climate research, aid and technology programs. *New York Times* (ClimateWire), February 14, n.p. Retrieved from http://www.nytimes.com/cwire/2011/02/14/14climatewire-republicans-gut-epa-climate-rules-slash-deep-87716.html?pagewanted=all. Accessed on December 30, 2013.

Muller, R. (2011, March 30). *Statement to the committee on science, space, and technology of the US house of representatives*. Retrieved from http://berkeleyearth.org/pdf/muller-testimony-31-march-2011.pdf. Accessed on December 30, 2013.

Oreskes, N., & Conway, E. (2010). *Merchants of doubt: How a handful of scientists obscured the truth on issues from tobacco smoke to global warming*. London: Bloomsbury Press.

Pew Research Center. (2011, December 1). *Modest rise in number saying there is "solid evidence" for global warming*. Retrieved from http://www.people-press.org/2011/12/01/modest-rise-in-number-saying-there-is-solid-evidence-of-global-warming/?src=prc-headline. Accessed on December 30, 2013.

Rahm, D. (2010). *Climate change politics in the United States: The science, the politics, and the prospects for change*. London: McFarland.

Roberts, J. T., & Kincaid, G. (2012). No talk but some walk: Obama administration rhetoric on climate change and international climate spending. Paper presented at the Annual Meeting of the American Sociological Association, Denver.

Romney, M. (2012). *Energy*. Retrieved from http://www.mittromney.com/issues/energy. Accessed on December 30, 2013.

Rose, N. (1999). *Powers of freedom: Reframing political thought*. Cambridge: Cambridge University Press.

Semeniuk, I., Tollefson, J., Wadman, M., Mann, A., & Dickey Zakaib, G. (2011). Obama resists research cuts. *Nature*, *470*, 313–315.

Strupp, J. (2013). Business journalists: Climate change deniers have no place in our reporting. *Media Matters*, July 3. Retrieved from http://mediamatters.org/blog/2013/07/03/business-journalists-climate-change-deniers-hav/194736. Accessed on December 30, 2013.

Tillerson, R. (2012, June 27). The new North American energy paradigm: Reshaping the future. Council of Foreign Relations. Retrieved from http://www.cfr.org/united-states/new-north-american-energy-paradigm-reshaping-future/p28630. Accessed on December 30, 2013.

Trenberth, K. (2012). Check with climate scientists for views on climate. *Wall Street Journal*, February 1, p. A14.

US Department of Transportation. (2012). *Adaptation planning*. Retrieved from http://climate.dot.gov/impacts-adaptations/planning.html. Accessed on December 30, 2013.

US House of Representatives. (2011, October 14). Discretionary Spending Recommendations Offered by the Following Members of the House Committee on Science, Space, and Technology: Rep. Ralph M. Hall, Rep. F. James Sensenbrenner, Rep. Lamar Smith, Rep. Judy Biggert, Rep. W. Todd Akin, Rep. Michael McCaul, Rep. Steven Palazzo, Rep. Andy Harris, and Rep. Randy Hultgren. Retrieved from http://science.house.gov/sites/republicans.science.house.gov/files/Letter_to_Joint_Select_Committee.pdf. Accessed on December 30, 2013.

Van den Berg, A., & Janoski, T. (2005). Conflict theories in political sociology. In T. Janoski, R. Alford, A. Hicks, & M. Schwartz (Eds.), *The handbook of political sociology* (pp. 72–95). Cambridge: Cambridge University Press.

Vastag, B. (2011). Congress kills request for national climate service. *Washington Post*, November 21, p. A3.

PART II
INSTITUTIONAL LOGICS

THE CULTURAL ROLE OF SCIENCE IN POLICY IMPLEMENTATION: VOLUNTARY SELF-REGULATION IN THE UK BUILDING SECTOR

Libby Schweber

ABSTRACT

The UK government's support for sustainable construction involves an explicit attempt to introduce a new institutional logic into the construction sector, while the use of Building Research Environmental Assessment Method (BREEAM) as a preferred policy mechanism exemplifies neoliberal use of voluntary self-regulation to promote policy goals. This paper uses the case of BREEAM to examine the role of science and scientific expertise in the exercise of neoliberal governance. More specifically, it combines a neo-institutional analysis of change with Foucault's theory of governmentality to explore the effect of BREEAM on eight construction projects. The concepts of visibility, knowledge, techniques, and identity provide an analytic grid to explore the effect of BREEAM on understandings and practices of "green building." Appeals

Fields of Knowledge: Science, Politics and Publics in the Neoliberal Age
Political Power and Social Theory, Volume 27, 157–191
Copyright © 2014 by Emerald Group Publishing Limited
All rights of reproduction in any form reserved
ISSN: 0198-8719/doi:10.1108/S0198-871920140000027014

to science and scientific authority are found to be most important in those instances where institutional logics clash and the legitimacy of BREEAM as a carrier of sustainable construction is challenged. From a theoretical perspective, the case studies highlight the role of instruments in the micro-dynamics of institutionalization. Empirically, it underlines the limited, but nonetheless significant, effect of weakly institutionalized neoliberal policy mechanisms.

INTRODUCTION

The introduction of sustainable development as a policy goal is generally ascribed to the Bruntland Report (U.N. World Commission on Environment and Development, 1987). Since then, a myriad of different definitions and associated goals have been offered at international, national, and regional levels. Regardless of how sustainability is defined, the aim is consistently depicted as a collective (inter-generational, transnational) problem, which cannot be adequately addressed by short-term market logics and mechanisms. Despite this, neoliberal states continue to privilege market-based tools such as carbon trading (Callon, 2009; Lohman, 2009; MacKenzie, 2009), stream mitigation banking (Lave, this volume) and voluntary standards (Potoski & Prakash, 2004, 2005; Short & Toffel, 2010). Based on science – or technical expertise of some kind – these tools tend to be presented as politically neutral. As such they point to one of the ways in which science is enrolled in projects of neoliberal governance. Within the construction sector building-level assessment methods such as the Building Research Establishment's Environmental Assessment Method (BREEAM) in the United Kingdom and Leadership in Energy and Environmental Design (LEED) in the United States epitomize this mode of governance. This paper takes the case of BREEAM to explore the role of science in the implementation of state policy for industry.

Viewed from the perspective of neoliberal governance, BREEAM exemplifies a type of voluntary self-regulation which neoliberal states champion. While much has been written on the formal features of BREEAM and comparable tools (Cole, 2005; Gibson, Hassan, Holtz, Tansey, & Whitelaw, 2005; Kajikawa, Inoue, & Goh, 2011) and while initial research has begun to explore their effect on market prices and buildings (Fuerst & McAllister, 2011; Lützkendorf & Lorenz, 2011), little thought has been given to these tools as coordinating mechanisms and what they teach us about the changing relations of science, industry, and the neoliberal state. Three features in particular remain relatively under-theorized: the role of science and expertise in their deployment, their effect on the design and construction of buildings (the production process rather than the product and its market value) and the insights which their development and use offers into neoliberal modes of governance and more specifically into the effect of weakly embedded governance tools. Most discussions of science and neoliberalism focus on policy formulation. In contrast, all of these issues touch on the effect of those policies on everyday practice, in this case on industry practice.

From the perspective of governance, the case of BREEAM offers an example of a weakly embedded instrument. It is weakly embedded in two senses. First, its aim is to adapt, rather than replace, dominant practices. Second, its use is not backed by strong legal or market sanctions. A premise of this paper is that the weaker the policy instrument (in the sense described herein), the greater the importance of science – or rather the cultural authority of science – in the (indirect) exercise of political authority. Thus whereas building professionals have no choice but to respect building regulations (at least in the United Kingdom, where inspections are frequent and regulations are enforced), they are less constrained when it comes to directives coming from voluntary policy tools. In this situation, issues of cognitive and normative legitimacy play a greater role in eliciting compliance, and science, as a source of cultural authority, figures more prominently.

From a theoretical perspective, these observations suggest the relevance of neo-institutional theory for exploring the role of science in the exercise of neoliberal governance in general and in the case of BREEAM in particular. A core issue for sociological neo-institutionalism (SNI) is the effect of field-level rules on firm-level behavior. The role of BREEAM in the (state's) explicit strategy for sustainable construction fits nicely into this model. However, as the discussion which follows indicates, it also draws attention to a number of gaps in that otherwise rich and coherent theoretical framework ("Theorizing Building Assessment Methods"). More specifically, it poses the problem of how to study the micro-dynamics of institutionalization. In terms of science and the state, it raises questions about the highly diffuse but nonetheless influential role of scientific expertise in the legitimation and exercise of neoliberal techniques of governance ("Findings"). These two discussions, in turn, support the introduction of Foucault's analytics of government as a conceptual framework from which to explore the cultural role of scientific authority in the deployment of neoliberal techniques of voluntary self-regulation in the construction sector ("Discussion"). The paper uses this lens to examine a number of cases of BREEAM assessment ("Conclusion") and to reflect on the role of science in the exercise of BREEAM and on the implications of the analysis for an SNI approach to the study of the micro-dynamics of (neoliberal) institutionalization.

THEORIZING BUILDING ASSESSMENT METHODS

SNI and Green Construction

SNI is a rapidly growing, relatively coherent area of research which has come to dominate organizational studies.[1] Its contribution to the study of sustainable construction lies in its emphasis on the role of non- or extra-commercial drivers in shaping firm-level behavior (Meyer & Rowan, 1977). Whereas much of the literature on corporate sustainability (Linnenluecke & Griffiths, 2013) focuses on a combination of firm-level capabilities (resource dependency theories) and immediate stakeholder demands (stakeholder theory), SNI highlights the role of the state and society. As such it would seem well adapted to research into the relations between science, industry, and the neoliberal state. Curiously, neither science or neoliberalism figures centrally in the literature (for an exception, see Campbell & Pedersen, 2001).

At the core of the SNI model is an analytic distinction between organizational fields and firms or actors. Organizational fields are conceptualized as a set of rules and understandings which shape firm-level behavior and are analyzed in terms of the type of sanction through which that influence is exerted (coercive, normative, and cognitive) (Scott, 1995). With respect to the construction sector, relevant institutions include building regulations and planning requirements (coercive), shared understandings of "best practice" (normative), dominant business models, standard operating procedures, and societal assumptions about buildings and their features (cognitive) (Hoffman & Henn, 2008). As this list shows, many of these rules and norms are associated with specific organizations and groups. These include regulatory bodies, professional societies, clients, firm-level competitors, supply chains, manufacturers, and civil society groups with a stake in the built environment. Whereas in Bourdieu's (1989) version of field analysis, fields are highly abstract analytic constructs, constituted around a single type of capital (political, economic, scientific, etc.); in SNI fields are more heterogeneous and more concrete. They are made up of different types of groups and organizations and substantive rules. They are also primarily ideational constructs.

This model of fields and of field-firm relations has been further developed through the concept of "institutional logics" which draws attention to the more systematic link between rules and practices (Thornton, Ocasio, & Lounsbury, 2012). A key topic in this literature concerns the impact of multiple logics on firm-level activity (Deephouse & Suchman, 2008; Pache & Santos, 2010; Suchman, 1995; Wooten & Hoffman, 2008). Initially, SNI studies of institutional logics focused on differences between commercial and non-commercial logics, including aesthetic logics (DiMaggio, 1991) and professional logics (Thornton & Ocasio, 1999). More recently, the concept has been applied to variations in commercial logics (Lounsbury, 2007). Within this model, there is some ambiguity as to whether institutional logics exist independently of individual instances of their enactment (Greenwood, Micelotta, Raynard, Kodeih, & Lounsbury, 2011). On the one hand, they are seen as external to individual or firm-level interpretations, but on the other hand they can only be observed through their expression in enacted roles and practices. Either way, they tend to be discussed as field-level phenomena.

In the past few years, SNI has been particularly important in theorizing the impact of a growing political and societal concern with climate change, energy security, and sustainable development on industry (for a review, see Hoffman & Georg, 2012). Environmentalism is offered as an example of the introduction of a new institutional logic into a commercial field. Whereas much of the literature on institutional logics treats multiple logics as a transitional phenomenon, research into business and the environment is distinctive for its recognition of the permanence of multiple institutional logics (Hoffman, 1999, 2001).

Turning to the firm level, SNI has long been interested in variations in firm responses to institutional demands (Oliver, 1991); however, it is only recently that scholars in this tradition have begun to explore the way in which firms negotiate multiple institutional logics (Kraatz & Block, 2008). Early work on this topic highlighted the phenomenon of decoupling, whereby firms engage with a particular demand symbolically, but not substantively (Fiss & Zajac, 2006). In the area of sustainable construction this is referred to as "green-washing" (Greer & Bruno, 1996; Hamilton & Gioia, 2009). While this mechanism is clearly important in managing contradictory field-level demands, the attention which it has garnered obscures the myriad of more subtle ways by which multiple logics combine at the level of the firm and everyday practice.

Within SNI, the overlay of distinct logics has been explicitly addressed in work on institutional change. A major criticism of SNI has been its neglect of agency and change (Hardy & Maguire, 2008). In response, a number of authors have focused on the role of institutional entrepreneurs, and in particular professionals, in the development and introduction of new institutional logics (Battilana & Casciaro, 2012; Battilana, Leca, & Boxenbaum, 2009; Hardy & Maguire, 2008). At the expense of caricaturizing a rich and sophisticated literature, institutional entrepreneurs are often presented as heroic figures committed to the promotion of alternate rules and practices (logics). The ability of such individuals to formulate alternate institutional rules is explained by the existence of field-level professional niches, which allow for the development of alternate ideas and practices. The introduction of these new institutional logics at firm level has been explained by the movement of these individuals between field-level niches and firm-level positions (Hoffman, 2001; Zilber, 2002). Recently, attempts have been made to link their effectiveness to structural features of the firm, such as the shape of informal networks (Battilana & Casciaro, 2012).

Instruments and the Micro-Dynamics of Institutionalization

The case of BREEAM and its effect on construction practice points to a number of under-developed areas in SNI theorizing. First, it poses questions about the role of techniques and instruments in the operationalization of ideas into field-level rules. As a number of commentators have noted, the theory currently overlooks the materiality of institutions and institutionalization (Czarniawska, 2008; Friedland, 2012; Gilad, 2012; Hasselbladh & Kallinikos, 2000). An important exception can be found in SNI work on formal standards. This literature uses SNI to draw attention to the extensive organizational activity involved in formulating and maintaining field-level rules – as exemplified by the production of ISO standards (Boiral, 2012; Sandholtz, 2012) and the FTSE4Good (Slager, Gond, & Moon, 2012). In contrast, the case of building assessment methods draws attention to institutional work at the firm or production level.

These observations highlight a second point regarding SNI, namely, its neglect of micro-level dynamics. While founding texts in SNI pointed to the importance of the micro-level internalization (Meyer & Rowan, 1977), the literature on institutional work has largely neglected these processes

(Lawrence, Suddaby, & Leca, 2009), preferring to focus on field-level activities. The discussion which follows makes a case for the central role of micro-level experiments and activities in the association of new ideas with specific practices and their adoption as taken-for-granted rules (cognitive institutions). The theoretical challenge of how to incorporate micro-level dynamics into SNI has been addressed from a number of angles. These include the combination of SNI with: ANT (Czarniawska, 2009; Tryggestad & Georg, 2011), creative action theory (Weik, 2012), practice theories (Lounsbury & Crumley, 2007; Whittington, 2006), identity theory (Kraatz & Block, 2008), and discourse analysis (Phillips, Lawrence, & Hardy, 2004), to name but a few. This paper follows Hasselbladh and Kallinikos (2000) in their use of a Foucauldian perspective.

The choice of Foucault's theory of governmentality and more specifically his analytics of government rests on the attention which Foucault pays to techniques and on his theorization of specifically neoliberal modes of governance. Combining an SNI model of field-firm relations with governmentality draws attention to the role of instruments in the operationalization of field-level ideas into firm-level practices.

A second point which the case of BREEAM raises for SNI concerns the specification of the relevant field. On the one hand, SNI defines a field as a comprehensive space encompassing the set of organizations and rules governing a particular type of activity. On the other hand, SNI also insists that the concept of field is an analytic construct which can and should be applied differently, depending on the research problem to hand. What counts as a field in one study might be part of a field in another. In the case of BREEAM and its effect on the construction sector, the question arises: What is the relevant field? Is it that of sustainable construction or of construction *tout court*? For the moment, the tendency is to adopt the first option (Boxenbaum, Georg, Rieijonen, & Garza De Linde, 2013; Hoffman, 2001); however, the identification of the ongoing negotiation of meaning and practices on the ground as institutional work challenges this position. It suggests that the specification of a new set of ideas and practices (institutional logic) is shaped by existing ideas and practices. Within SNI, this problem has been alluded to in discussions of the fit of new institutions with the status quo (Amis, Slack, & Hinings, 2004; D'Aunno, Succi, & Alexander, 2000; Greenwood & Hinings, 1996). This paper contributes to that problematic by focusing on the role of instruments in the inflection of dominant practices. More specifically, it looks at the impact of BREEAM on the construction field *writ* large (rather than sustainable construction).

This triple concern with micro-level dynamics, instruments, and underspecified institutional rules leads to a reformulation of the process of institutional work and institutionalization more generally. Instead of treating the process as an activity which takes place at the field level, independent of individual firms and actors, this paper explores it as an emergent process, in which techniques and instruments figure as the focus of ongoing interpretation and negotiation. Viewed thusly, institutionalization becomes a problem of how to account for the emergence of agreed-upon meanings and practices which, once stabilized, take on an existence and force of their own, independent of the myriad of micro-level experiments which produced them. The case of BREEAM provides an opportunity to examine that process prior to stabilization.

This problematization, in turn, opens the way for a very different type of inquiry into the use of science in neoliberal governance of industry. An important contribution of SNI lies in the observation that firm-level engagement with external rules – be they legally binding regulations or social norms and expectations – is a meaning laden process (Hallett & Ventresca, 2006; Meyer, Boli, & Thomas, 1987; Weik, 2012) informed by concerns with legitimacy (Meyer & Rowan, 1977). For the most part, legitimacy is seen to depend on conformity with already established and known practices (associated with ideas – expressed in norms or rules). The case of BREEAM provides an opportunity to examine the role of science and technical expertise in that process.

The Cultural Authority of Science

The relation between science and the state can be approached from a number of different perspectives. This paper takes as its starting point Yaron Ezrahi's (1990) work on the role of science as a political resource in Western liberal democracy and Theodore Porter's (1995) analysis of the role of quantification in democratic states. The relevance of these two scholars and their work for this paper lies in the emphasis which they place on science as a cultural resource and source of legitimacy in liberal democratic states.

In *The Fall of Icharus* (1990), Ezrahi argues that in the first two thirds of the twentieth century, science played a central role in the legitimation and exercise of political power in the United States and England. In support of this claim, he points to the importance of observation as a basis of scientific and, by extension, political authority. Public debate, voting, and representative assemblies draw on a scientifically inspired model of how to arrive at depersonalized, universal truths. Porter (1995) makes a parallel argument about quantification and its use to depersonalize policy making and inspire trust. According to Porter, quantification and democratization were historically linked, through the substitution of supposedly universal numeric standards for political decisions. As both authors argue, in the twentieth century "science" – or rather the "scientific method" – were depicted as representatives of "the public interest" and used to legitimate state intervention in civil society and the economy.

In his book, Ezrahi goes on to argue that since the 1970s the cultural role of science in liberal democracy has declined with the postmodern public loss of faith in science. This shift, he argues, can be seen in the depiction of science as an economic resource rather than an intrinsically valuable cultural activity (1990, p. 278). But this does not mean that science has disappeared from politics, nor that it has ceased to play an important cultural role. Instead, as a growing number of studies show, science has increasingly become a locus of political contestation (Epstein, 1998; Kinchy, 2012; Kleinman, Kinchy, & Handelsman, 2005). These studies all focus on political activism and policy formulation. In contrast, the problem of institutionalization and the case of BREEAM opens a new area of research, namely, the role of science and scientific authority in the implementation of neoliberal policies.

The emergence and character of the neoliberal state is discussed below. At this point in the discussion I want to signal the neglect of science in these discussions, especially when it comes to state-industry relations. A central argument in the literature on neoliberalism involves a change in the nature of state involvement in "society," as marked by the shift from a command-and-control approach to a more managerial role. While much has been written on neoliberal tools of governance, less has been said about the role of science as a cultural resource in the legitimation and deployment of those tools. Two exceptions include recent work in the new political sociology of science (NPSS) and Foucauldian studies of public sector activities, such as health and education.

For the moment, much of the work in NPSS focuses on the role of scientific discourse and expertise in the regulation of technology (Moore, Kleinman, Hess, & Frickel, 2011). This paper extends that problematic to the regulation or rather management of economic activity. It also shifts the focus of attention from policy formulation to policy implementation. Thus, whereas much of the NPSS literature focuses on the mobilization of science and scientific criteria in debates over specific laws or directives (Hess, 2007,

2011; Kinchy, Kleinman, & Autry, 2008; Kleinman, Kinchy, & Autry, 2009), the case of BREEAM draws attention to the cultural role of science in contests over the application of technical expertise.

A second exception to the neglect of science in the literature on neoliberal politics can be found in Foucauldian studies of public sector policies (Cotoi, 2011; Dean, 2007; Newman, 2005). A central argument in much of this work involves the role of social sciences in the development and legitimation of indirect modes of government. As the discussion which follows indicates, this literature has the advantage (for my paper) of examining policy implementation. Curiously, in the shift from policy formulation to implementation, the cultural force of science largely disappears from the analysis. Science, in these accounts, remains wholly on the side of the governing. Thus while teachers and public health workers do resist the growing and (in their view often) distortive effects of targets and benchmarks, their resistance is rarely framed in scientific terms. Instead it is usually depicted as a contest between managerial and professional, humanist, or aesthetic values (Ball, 2013; Prozorov, 2007). As the discussion of BREEAM below indicates, this issue of science and managerial values figures in construction projects as well. However, in this instance scientific and engineering expertise is distributed across the certifying bodies, assessors, and subjects of the assessment, thus providing for a very different dynamic.

As this discussion suggests, the study of the effect of BREEAM on construction activity offers a very different kind of case for the study of science, industry, and the neoliberal state, one where policies and tools are developed and legitimated in scientific and technical terms and in which user engagement is contingent, at least in part, on acceptance of those claims. As such, the case provides an opportunity to extend arguments concerning the cultural authority of science in neoliberal politics. Relevant questions concern the extent to which various publics accept or challenge the technical basis of such tools in the course of their everyday use and the effect of this contest on their engagement with them.

Foucault, Neoliberalism, and Techniques of Governance

For many political observers, the economic crises of the 1970s and 1980s mark a break in state-society relations in Western liberal democracies and more specifically in Anglophone countries, notably the United States and the United Kingdom. Whereas for most of the twentieth century, liberal democratic or welfare state models dominated political discourse, since the election of Reagan in the United States and Thatcher in the United Kingdom a new neoliberal ideology has taken their place (Fourcade, 2009; Gamble, 1988). At the center of this development is an ideological belief in the "necessity and desirability of transferring economic power and control from governments to private markets" (Centeno & Cohen, 2012). This principle has consequences both for the relation of the state and society and for the basis of the political contract. In principle, it supports market deregulation, state decentralization, and reduced state intervention in economic affairs (Campbell & Pedersen, 2001). Whereas liberal democracy rested on the state's commitment to promote the "public interest," neoliberal states see their role as protecting values of personal engagement, responsibility, transparency, and choice (Newman, 2005; Pestre, 2008).

On the surface, this ideology and the policies which it supports would seem to leave little room for science or the state. However, underneath – or rather, supporting – these ideological commitments are a panoply of logistical ordering devices (Freiburger, Biggart, & Beamish, 2012) by which the state supports "free markets," opening the way for often extensive, albeit indirect, state intervention. This activity is clearly evident in the role of the state in the specification of commodities and value, as well as the regulation of the financial system. Foucault's work on governmentality explicitly theorizes this "control of control."

In an early lecture at the Collège de France, Foucault described a historic shift from direct state intervention in the lives of citizens to "control at a distance" (Lemke, 2001). In this new mode of governance, coercive state power is increasingly replaced (or some would say complemented) by indirect mechanisms of control (Dean, 1999; Gane, 2012; Miller & Rose, 1990; Rose & Miller, 1992). The state's increasing reliance on mechanisms such as standard setting and voluntary self-regulation has been explored under the heading of the "regulatory state." As Moran explains, the EU and modern (neoliberal) state is regulatory in two senses:

> It largely confines itself to the creation of frameworks of rules that are then implemented elsewhere; and it practices regulatory policies – addressing the task of remedying market failures rather than the more ambitious interventionism of the Keynesian era. (2003, p. 6)

An important tool in the neoliberal arsenal is the use of auditing techniques to establish norms and control behavior (Power, 1999). Viewed from this perspective, one could argue that the observation of political decision making which Ezrahi analyzed has been replaced by the observation (and constitution) of subjects. Both modes, it should be noted, rest their claims to legitimacy on the cultural authority of science. Whereas for both Ezrahi and Porter, science was used to hold politicians to account (or at least to be seen to be doing so), in Foucault's version it is used to define problems, specify practices and discipline subjects.

According to a number of authors, this form of neoliberal governmentality has developed the most intensively in the United Kingdom (Moran, 2003; Power, 1999). Since the 1980s the United Kingdom has undergone a radical program of privatization and deregulation. In contrast to the United States, in the United Kingdom neoliberalism has involved a dramatic centralization of political power and authority and an associated increase in the micro-management of both public and private sector activity from the center.

The effect of these developments on the construction sector stems from the state's dual role as (neoliberal) manager and client. The election of the Labor Government in 1997 was accompanied by the development of a concerted attempt by central government to "improve" the construction sector. The reform agenda was laid out in documents such as the Latham Report, *Constructing the Team* (1994) and the Egan Report, *Rethinking Construction* (1998) and accompanied by a frenzy of managerial activity. In terms of vision, this reform movement was driven by a concern to apply lessons from the aeronautics and petrochemical sector to the construction sector; more specifically it involved the application of techniques of new public management and total quality control (Green, 2011).

Two aspects of this improvement agenda are relevant for this paper. The first involves the familiarity of construction-sector professionals with neoliberal management tools. Far from a novel type of instrument, BREEAM is just one more in a long list of management techniques and devices to which construction professionals are subject. The second concerns the tension between general targets and local conditions inherent in this mode of governance. A major criticism of neoliberal policy in general and the improvement agenda in particular is its failure to take into account the specificity of the sector and the impact which universal management fixes might have (Green, 2011). As the discussion below demonstrates, the same issue emerges at a more micro-level every time such tools are implemented. A central question for this paper concerns the dynamics which this tension generates and the role of science and technical expertise therein.

Turning to the state's role as client, neoliberal reform dramatically changed the market for the construction sector. Privatization of key industries deprived the sector of a predictable and steady source of infrastructure and public sector demand; at the same time, centralization and the government's explicit use of its role as client to influence industry behavior introduced a wide variety of sector level targets and benchmarks. The uptake of BREEAM since 2008 can be directly linked to internal compliance targets for government procurement and its identification of BREEAM as a preferred policy mechanism for the promotion of sustainable construction. Since 2008, new public sector building projects require some form of building-level environmental assessment.

Returning to the problem of the relation of science, industry, and the neoliberal state, Foucault's concept of governmentality helps to (re-)formulate the question of the cultural authority of science. It calls for an expansion of the problematic from policy formulation to policy implementation. Just as research into the institutionalization of new rules needs to be extended beyond the field level to the firm level, so too inquiry into the role of science needs to go beyond formal policy debates to far more diffuse and distributed effects at the level of everyday practice. Whereas for policy makers disciplinary divisions may matter, at the level of firms and everyday work practices, the distinction between scientific and technical knowledge is less relevant. For the purposes of this paper, the cultural role of science will be used to refer to claims made in the name of science and engineering expertise.

A central research question concerns the conditions under which neoliberal instruments are or are not adopted or challenged and the place of science therein. When it comes to BREEAM, the legitimacy of the technique clearly depends on some vague association of science, technical expertise, quantification, universalism, and depersonalization, as well as a perceived relation between the concept of green building and the rules and practices which BREEAM as an instrument supports. But how effective is that claim and does it matter? Or rather, under what conditions does it matter? Does confidence in the scientific basis of standards and benchmarks influence their uptake and use? Under what conditions do scientific claims become the object of contestation and in what type of instances are they taken for granted?

BREEAM and the Analytics of Government

As indicated above, Foucault's work on neoliberal forms of governance (Foucault, 1977, 1991[1978], 2008) has been crucial in underlining the seemingly paradoxical relation between an ideology of deregulation on the one hand and the centralization and intensification of power on the other hand. At the core of Foucault's approach is a shift in the focus of political analysis from a study of the state to a study of governmentality (Dean, 1999). Whereas "the state" generally refers to a set of official bodies and coercive activities, "governmentality" refers to any attempt to control and direct behavior. It includes formal and informal mechanisms, their enactment and the ways of thinking and acting which inform them.

Drawing on Dean's application of Foucault's concepts to the study of neoliberalism, Gouldson and Bebbington (2007) suggest that this framework can be used to study environmental governance, of which BREEAM is a component. More specifically, they distinguish between three analytic components: problematization, analytics of government, and utopian ideal. Of the three it is analytics of government or the mechanisms by which neoliberalism is enacted in everyday life which provides the focus of this paper. Whereas most empirical studies focus on the tight hold which governmentality exercises over individuals, the focus of this study is on the conditions for its exercise. More specifically, the analysis explores the role of science as a cultural resource in the legitimation of specific instances of governmentality.

The analytics of government identifies a number of mechanisms which serve to embed institutional logics. These include visibilities, knowledge, techniques and practices, and identities (Gouldson & Bebbington, 2007). Applied to the study of environmental assessment, these categories draw attention to the problems and issues rendered visible in the course of an assessment process, to the knowledge generated in the development of BREEAM and its application to particular building projects, to the techniques and practices mobilized in its implementation, and to the effect of the assessment process on the identities of the many actors engaged in its construction. Reframed in SNI terms, the analytics of government offers a framework to explore the elaboration of new institutional logics through the linkage of new ideas (in this case green building) with specific practices at the level of everyday practice.

Foucault's analysis of the role of techniques and knowledge can be used to explore the role of scientific authority in the legitimation of neoliberal governance at the level of practice. In Foucault's account of governmentality, these "scientifically" grounded techniques work by establishing norms which subjects then internalize. For the researcher this poses a problem, as the effect of science is largely invisible, except when norms and rules are challenged. The discussion of findings which follows explores this suggestion. More specifically, it uses the process of a BREEAM assessment to explore bases for subject level acceptance of and challenge to the association of the concept of "green building" with a particular set of rules.

Building on Foucauldian studies of audit and assessment, the research begins from a model of BREEAM as a process which is super-imposed on the design and construction processes. Data were collected between August 2010 and December 2011, using a combination of semi-structured interviews and documents produced by the assessment process. For each case, 1–2 hour interviews were conducted with the main actors directly involved in the assessment process. In terms of formal roles this included some combination of the BREEAM assessor, client, architect, project manager, design manager, and specialist engineers (usually the mechanical, electrical, and structural engineer). In total 49 interviews were conducted. Questions focused on project team members' experience in the assessment process, its effect on their everyday work, and its effect on their understanding of BREEAM, green building, and sustainable construction. Data were analyzed using the analytics of government framework described above. Ethics permission was secured from the University of Reading and professional guidelines respected. The discussion which follows reports on those aspects of the analysis which shed light on the use of science, or rather appeals to scientific authority. For a more detailed account of the findings see Schweber (2013) and Schweber and Haroglu (2014).

FINDINGS

BREEAM as a Neoliberal Technique

BREEAM is a comprehensive, multi-criteria assessment method. The method is the creation of the Building Research Establishment (BRE), a nonprofit private foundation devoted to research on the construction sector (Courtney, 1997). The organization began its life as a government-based scientific research institute into the performance of building materials. During the 1970s, it increasingly assumed the role of scientific advisory body to the government on a range of construction related matters. With the state's turn to neoliberalism in the late 1980s, it was transformed, first into a quasi-independent executive agency and eventually into a private nonprofit foundation. In the process, the newly created BRE Trust was allowed to keep BREEAM, thus providing it with a significant edge in the market for environmental building methods (Cole, 2006). At the moment,

BREEAM is one of the BRE's primary sources of revenue. This includes income from the training of BREEAM assessors and certification of buildings. Thus, while the BRE presents BREEAM as a scientifically informed, disinterested instrument, the method, its deployment, and marketing are vulnerable to concerns regarding commercial interests.

The method itself consists of a framework of nine categories, each corresponding to a different aspect of green building: energy, transport, water, materials, waste, land use, and ecology, pollution, management, and health and well-being. Each category is broken down into a variety of performance criteria and other measures, each of which is assigned a certain number of credits. The method calls upon design and construction teams to adopt a wide range of new practices and, in some cases, to inflect existing ones. These range from introducing new types of materials and technologies (such as recycled aggregates for concrete, automatic sanitary shut outs in bathrooms and low carbon technologies), to new practices (such as refrigerant leak detection and external lighting levels), to scoring well on other instruments (such as Energy Performance Certificates (EPC) or flood risk surveys).

To complete an assessment, members of the project team produce evidence to establish their compliance with these criteria. These are compiled by the BREEAM assessor and sent off to BRE Global Ltd., which evaluates the application and comes back with a score. Scores are calculated on the basis of the accumulated number of credits, weighted by category. These scores are then used to assign the building a rating, ranging from Poor to Outstanding. There are currently 16 different versions of BREEAM for different types of buildings, including BREEAM Offices, BREEAM Retail, and BREEAM Education. In addition, each version is updated every couple of years (Atkinson, 2009), in part to stay ahead of changing building regulations.

In principle, the use of BREEAM is voluntary; in practice, its use is encouraged by the state. Introduced in 1990, BREEAM was the first building-level environmental assessment method to be developed. Within the United Kingdom, uptake was initially very slow. In 2000, BREEAM began to be included in the list of strategic tools which UK policy makers recommended for use. In 2000, the government also adopted BREEAM as a mandatory mechanism for all government procurement (DEFRA, 2005; OGC, 2008). At the same time, local authorities began to include BREEAM in their newly required sustainable development strategies. The result is that, while large segments of the construction sector can and do ignore BREEAM, those firms involved in publicly funded buildings and those with sustainability minded clients have had to accept the overlay of a BREEAM assessment process onto the design and construction process. By 2013, over 15,000 projects had been BREEAM certified, with half dating from 2008 and onwards (BRE Global, 2013a).

The case of BREEAM is of interest in that it offers an example of an explicit attempt to shape everyday practice by establishing voluntary "best practice" rules (Kerwer, 2005) and embedding them in a market mechanism. In considering the role of standards in shaping everyday behavior it is useful to distinguish between formal and informal standards. Theoretical reflection on formal standards tends to focus on their use in transnational governance. Formal standards are rules which are developed to coordinate organizational activity (Botzem & Dobusch, 2012). They tend to share a number of characteristics, including their production by non-state organizations, an appeal to experts and scientific authority, pressure for compliance from third parties (Brunsson, Andreas, Rasche, & Seidl, 2012), and a tension between the general standard and the specificity of object to which it is being applied (Timmermans & Epstein, 2010). Informal standards, in contrast, refer to taken-for-granted understandings of standard and best practice. As such, they correspond with the SNI concept of cognitive institutions, which specify rules and practices.

BREEAM shares many of the features of formal standards. Its aim is to set a standard for green building and, in so doing, to inflect design and production decisions. Like transnational standards, it is owned and managed by an independent body of technical and scientific experts, with claims to political neutrality. In principle, BREEAM ratings were designed to provide an indicator of quality and thus market value. In practice, the BRE has yet to realize this aspiration.

The BRE website lists "market recognition" as one of the reasons to perform an assessment (BRE Global, 2013b); however, the evidence for this claim remains ambiguous. Studies seeking to establish the financial returns on BREEAM assessments and other types of environmental certification suggest a very limited impact (Fuerst & McAllister, 2011; Lützkendorf & Lorenz, 2011). These findings are confirmed by the research reported herein. Amongst the 49 construction professionals interviewed (including clients), none of them cited market value as a reason for seeking BREEAM certification. Instead, reasons for conducting an assessment included client requirements (and in particular public sector clients), planning conditions, reputational value, and sustainability.

These observations serve to establish BREEAM as a neoliberal mechanism whereby the state mobilizes instruments produced by independent "scientific" or rather expert organizations to promote the public interest. The absence of strong coercion or market drivers suggests the importance of normative and cognitive legitimacy amongst construction professionals who participate in the assessment process. The discussion which follows draws on the Foucauldian categories of visibilities, knowledge, techniques and practices and identity to examine the dynamics which a BREEAM assessment generates on the ground.

Visibilities

The concept of visibilities draws attention to the way in which BREEAM as an instrument renders visible certain aspects of a green building and obscures others. Stated differently, it points to the role of the instrument in the specification of a field-level idea. More importantly, it highlights the role of instruments in linking new ideas to rules (directives for practice). In the case of BREEAM, these effects stem both from the formal features of the instrument and its application.

In terms of formal features, BREEAM explicitly privileges environmental over social and economic dimensions of sustainability and measurable criteria over less quantifiable features (Du Plessis & Cole, 2011). This bias is embedded in the weightings which favor environmental categories (energy, water, materials, waste, land use, ecology, and pollution) over social ones (management, health and well-being, and transport) and in the specification of credits which favor quantifiable measures of performance.

One of the key features of governmentality concerns the way in which instruments actively specify broader policy goals. In the case of BREEAM, this effect can be seen in the use of metrics to define otherwise open or fluid concepts. To give but a few examples, BREEAM defines "biodiversity" as six species or more. It defines "proximity to amenities" as "a situation in which the main entrance to the building is located with 500 meters of a grocery shop/food outlet, post box, or cash machine, provided they can be accessed by a 'safe pedestrian route.'" It limits "community engagement" to formal meetings for which there was a consultation plan identifying "points at which consultees could usefully contribute and how they would be informed about the progress of the project." And it excludes embodied carbon from the "reduction of carbon dioxide emissions." In each case, the instrument serves to operationalize an otherwise underspecified – and in some instances contested – concept.

From the perspective of project-level stakeholders the issue of visibility draws attention to two features of their engagement with BREEAM. These include first, the role of the assessment process in rendering visible particular credits and issues and second, the opportunity which it creates for explicit reflection and challenge. When it comes to the first, a number of interviewees commented on how participation in the assessment process had introduced them to a range of issues which were outside their usual range of vision. This effect resulted from the requirement that project team members attend BREEAM meetings in which progress on each credit was systematically reviewed. At the end of the assessment process, professionals can be expected to have been alerted to a set of topics which, taken together, define the concept of "green building." But to what extent do they buy into this operationalization?

Participation in the BREEAM process both introduces a particular definition of "green building" and provides an opportunity for reflexive engagement with it. The extension of the BREEAM process over a period of months and in some cases years means that individual professionals have numerous opportunities to discuss credits and their reasonableness with their colleagues within the team. Over and over in the course of the interviews, construction professionals challenged particular requirements as non-sensical. Not surprisingly, the more actors know about a topic or issue, the less likely they were to buy into and internalize the BREEAM specification. In the course of the interviews, subjects challenged specific credit definitions, the relative weightings (or number of credits) assigned to specific categories, and the "fit" of BREEAM with genuine thermal comfort or community engagement. A similar issue arose as to whether BREEAM adequately captured the idea of "green building." For a number of interviewees, the instrument was good at what it did but inadequate as an operationalization of genuine sustainability. Specific observations and criticisms aside, the fact that construction professionals felt empowered to actively engage with and pass judgment on specific requirements draws attention to the distribution of scientific or technical authority across stakeholders in the assessment process.

One issue which arises in this context is the extent to which the involvement of the entire project team in BREEAM meetings "protects" against the black boxing of the instrument. Many of the construction professionals described the arbitrariness of particular specifications, while others explicitly criticized particular credits or categories for not adequately capturing the concept at hand. In most cases, this skepticism as to the trustworthiness of the numbers was not enough to overrule their engagement with the instrument, but it did challenge the authority of both the BRE and the instrument. This point, in turn, highlights a contrast between construction professionals and the broader public whose only encounter with BREEAM is a final building rating of Very Good or Excellent and who have little to no understanding of what is behind the rating (although BREEAM does award a credit for communicating environmental features of the building to its users).

Viewed from the perspective of SNI, these micro-level challenges to the specification of specific credits points to a perceived disjuncture between the idea on the one hand and the instrument and associated set of practices on the other. It also highlights a distribution of expertise, such that project-level stakeholders see themselves as equally, if not more, qualified to specify what counts as a green building and to challenge BRE decisions.

Knowledge

For Foucault and Dean, the category of "knowledge" refers to knowledge regimes rather than information. The concept draws attention to the different types of reasoning at play and to the way in which they affect behavior (Dean, 1999, p. 31). When it comes to BREEAM, the concept raises questions about the different modes of reasoning and criteria of evaluation at play in the assessment process.

In general, construction work involves a combination of commercial, professional, technical, bureaucratic/administrative, and practical/problem-solving modes of reasoning (to name but a few). The BREEAM assessment process also combines these different types of knowledge, but it does so with different criteria or aims in mind. Thus, whereas traditional construction practices focus on delivering projects at cost and on time, BREEAM (in principle) calls for attention to the long term quality and sustainability of the building. A central question for this paper concerns the place of scientific and technical reasoning relative to other types of knowledge within the assessment process.

While it is not explicitly stated, the BRE's claims for BREEAM rest on the scientific and technical expertise which the tool embodies. At the same time, the BRE knows that for BREEAM to be adopted, the method must also be commercially viable and practical. The result is that BREEAM assessments are always informed by both technical and commercial considerations. A second, and perhaps more important, knowledge dynamic which informs the assessment process is between substantive and procedural or bureaucratic requirements. This concern takes two distinct forms. The first involves a tension between "genuinely" sustainable actions (be it a design decision or a process) and actions taken purely to gain credits. The second concerns the difference between assessment related activities which directly affect core building processes (design, procurement, construction, etc.) and those devoted purely to evidence gathering.

In terms of the first, a number of interviewees expressed their surprise that features which they deemed to be sustainable were not recognized within BREEAM. Similarly, they noted instances in which features which did not enhance the sustainability of the building were rewarded with credits. For example, professionals criticized the failure to give credits for composite materials, which combined recyclable materials with small amounts of non recyclable materials, be it in the structure, windows, or flooring. They also commented on the disproportionate number of credits given to seemingly insignificant features such as the location of a water fountain or coverings for cycle sheds. Finally, they noted the inclusion of requirements which were impossible to achieve, such as certain lighting specifications.

In each of these instances, the "knowledge" issue (in Foucault's sense) is not specific scientific or technical claims, but rather their operationalization into practical guidelines and requirements. And the implicit challenge is not to technical modes of reasoning (which construction professionals share) but to the BRE's application of those principles to building-level requirements. When it comes to BREEAM, the BRE is generally respected for bringing scientific knowledge about buildings and environmental science together in a comprehensive framework, but it is criticized (in particular instances) for not adequately translating that knowledge into practice. In terms of knowledge, it is not the type of knowledge that is in question but the ways in which multiple types are combined.

Turning to the second issue, at the core of the BREEAM assessment process is a tracker sheet which the BREEAM assessor prepares and which assigns responsibility for specific credits to particular individuals. To meet this responsibility the relevant professional needs both to ensure that the relevant design feature is incorporated into the design or building and to provide evidence that this has been accomplished. This evidence is included in the formal application for certification which the BREEAM assessor sends to the BRE. For many construction professionals, a key difficulty with BREEAM involves the excessively rigid and bureaucratic character of the requirements for evidence. Thus, while many of them are on board with the principle of green building and accept the specification of credits and their requirements, they resent the additional work which it takes to produce evidence in precisely the format which the BRE requires. In these comments, the issue is not the science or technical expertise embedded in BREEAM but rather the rigidity and inflexibility of the BRE and an associated concern that bureaucratic logics are being allowed to trump sustainability considerations.

Techniques and Practices

This concern with evidence draws attention to the techniques which translate knowledge into specific directives for action and their impact on everyday practice. A key difference between governmentality and other forms of power (or management) involves the type of sanctions applied to instances of non-compliance. Whereas historically governance relied on coercion or formal sanctions, governmentality relies on moral pressure.

A striking feature of BREEAM is the reliance of the BREEAM assessor and whoever else is driving the assessment on moral persuasion and the absence of formal sanctions for non-compliance within the project team. One of the main challenges in conducting a BREEAM assessment is getting the different members of the project team to deliver what they promised at the initial assessment meeting. This task is made all the more difficult by the amount of unpaid work which BREEAM generates and the absence of clear ownership over the process. Thus, while BREEAM assessors are paid a fee for their services, none of the other project team members is renumerated. Nor is the fee sufficient to cover the daily follow up needed to monitor compliance with actions on the tracker sheet or to chase up missing evidence. The result is that the task falls to whoever cares the most or is most involved with the project as a whole, usually the design manager or project manager, although in some instances it may be the architect (Schweber, 2014).

In this situation, the question arises: why should team members comply and what is the effect of scientific or other forms of legitimacy on encouraging compliance? Reasons given in the course of the interviews included: the desire to produce a green building, a taken-for-granted culture of professionalism that involves completing the tasks to which one was assigned, a team commitment to deliver the project, and reputational value for the different participating firms. Of the four, only the first depended on the scientific authority of the method and consequent acceptance of the equation of BREEAM with "green building." While problems ensuring that certain tasks were met and that evidence was produced on time were frequently mentioned, at no point did anyone criticize a colleague for not complying and there were no obvious penalties. As for the knowledge dimension, while there is no hard evidence, it does seem clear that project team members tend to neglect, or at least to postpone, those actions which are at odds with their understanding of either the project or sustainability or both.

Turning to the impact of BREEAM on design and construction practice, one of the striking features of the research was the convergence of BREEAM requirements with taken-for-granted understandings of standard practice and best practice. In the course of the interviews, construction professionals were given a list of credits for which they were personally responsible and asked to comment on the tasks. In many instances, credits were dismissed as "standard practice." When asked if they would have done the same thing three or four years ago, most responded that they would not have done so, thus suggesting that BREEAM is either driving or reflecting ongoing changes in standard practice.

A central feature of formal standards in general and BREEAM in particular involves the relatively general character of the performance and prescriptive targets which they prescribe. The result is that BREEAM criteria often clash with local conditions and requirements, rendering otherwise reasonable requirements irrational and creating a sense of unfairness amongst project team measures. To give but a couple of examples, project team members complained of having to do a flood risk survey for a building perched high on a hill, of losing credits for not introducing internet access into a mental health facility in which patients were not allowed to communicate with the outside world and of the impossibility of introducing rain water harvesting because of the slope of the roof. Once again, these examples point to a challenge to the BRE's operationalization of abstract scientific principles into project-level requirements.

This sense of unfairness was exacerbated by confusion over who and what is being evaluated. In principle BREEAM offers an assessment of a building. However, because the assessment is carried out by construction professionals and because their primary motivation is reputational, they interpret it as a measure of their collective performance. This then generates a sense of unfairness when credits are "lost" owing to factors over which they have no control. This feeling was particularly common when it came to transportation and amenity credits which depend on the building's proximity to public transportation and shops.

One final point regarding BREEAM as a technique concerns its use as a resource to manage project team relations. Until now the discussion has focused on BREEAM as an external requirement imposed on the project team. While accurate, this depiction obscures the active use of BREEAM by project team members to further or defend their own interests. Construction projects involve multiple professionals and firms, each with their own interests, commitments, and understandings of their particular remit and of what counts as good work and green building. A common tension is between the architect and the contractor, who often bears to the risk of going over cost and thus has a tendency to alter designs so as to save money. One of the more interesting findings to emerge from the study concerns the way in which sustainability minded architects appealed to BREEAM requirements to defend their initial design from cost cutting measures on the ground.

Identity

The final category in Foucault's analytics of government is identity. In most studies of governmentality this is the primary mechanism for the exercise of coordination and control and thus the focus of attention. In the case of BREEAM and, I would argue, other types of voluntary self-regulation, the identity effects are less significant. The important questions for this paper are: What are these more attenuated identity effects? and Why does BREEAM have less effect on the constitution of subject than the theory would suggest?

As this discussion indicates, BREEAM involves the responsibilization of individual professionals for specific tasks, both technical and bureaucratic. According to Foucault, the effectiveness of this mechanism depends on the internalization of BREEAM as a norm which commands voluntary compliance. However, as the previous discussion indicates, many construction professionals reject the automatic equation of BREEAM with green building. Instead, they engage with the content of the method, reflecting on whether specific credits fit with their own understandings of the broader concept and evaluating BREEAM requirements in the context of specific projects and sites. To the extent that BREEAM falls short, they still conduct the assessment, but in a much more superficial way.

If sustainability minded professionals approach BREEAM with a healthy skepticism, what of other construction professionals with less personal commitment or expertise? On the one hand, the less commitment and technical knowledge interviewees had about a specific issue, the more likely they were to accept BREEAM's operationalization of the concept, be it green building or more specific topics, and the more likely they were to accord the process with scientific legitimacy. However, this scientific and technical legitimacy did not enhance their engagement with the assessment process. To the contrary, one of the main research findings in this study was that the less a project team shared an independent commitment to sustainability, the less likely BREEAM was to inflect core construction processes and the less it entered into professional identity or norms.

This weakness – relative to similar tools of classification, quantification, assessment, and ranking – is partly explained by the orthogonal relation of BREEAM to both construction professionals and professional practices. As mentioned above, BREEAM does not assess individuals or even project teams, but buildings and, as construction professionals know well, the building is always subject to events beyond their control. These include site level issues as well as endless contingencies which lead to an ongoing adjustment and modification of the initial design. This distance serves to attenuate professional identification with BREEAM and its score, rendering it a weaker, albeit not irrelevant, neoliberal mechanism.

DISCUSSION

An analysis of the mapping of the BREEAM assessment process onto the design and construction process using the analytics of government draws attention to an ongoing negotiation around the fit between the idea, instrument, and set of practices. The concept of visibilities underlines the role of instruments in the specification of new institutional ideas. In the case of BREEAM, the assessment process presents construction professionals with a clearly articulated, albeit complex, list of the practices which specify the concept of "green building." One effect of the assessment process is to communicate and affirm that definition. Paradoxically, a second effect is to create a venue for critical reflection on its adequacy. By bringing all of the project team members together, each with their different competencies and areas of expertise, and by collectively discussing the achievability of specific credits, BREEAM opens itself up to challenge.

The discussion of knowledge effects in BREEAM draws attention to the different bases on which these challenges are expressed and specific practices legitimated. As the NPSS has shown, when it comes to policy formulation, policies are generally supported by a distinct type of claim, be it arguments about scientific method or social justice. When it comes to policy implementation, in contrast, practices are always "overdetermined," combining scientific or technical claims, with practical, financial, bureaucratic, and other considerations. In this case, legitimation depends on the (perceived) hierarchy of different types of knowledge. Thus, in the case of BREEAM, construction professionals accepted as legitimate those elements (credits) in which science was seen to trump bureaucracy; they rejected those elements in which bureaucratic or audit considerations seemed to prime. In both instances, their criticisms focused not on the type of knowledge at play, but on the inflexibility of the BRE in translating substantive, science-based requirements into specific practices and associated evidence trails. Viewed from this perspective, scientific authority was critical in legitimating the imposition of otherwise illegitimate bureaucratic or audit demands.

Turning to the effect of BREEAM as a technique, the analysis underlined both its role in linking ideas to specific practices and its weakness. At a very practical level, the assessment method translates the concept of green building into a myriad of discrete concepts – such as natural ventilation, thermal comfort, and biodiversity – and links each one to a discrete set of actions (and credits). This act of assemblage, I would argue, is an essential aspect of institutional work. At the same time, BREEAM on its own is too weak to fully institutionalize its own construction. In contrast to other neoliberal techniques, BREEAM has relatively little hold over its users. Thus, while the government sets BREEAM targets for new hospitals and schools, there are no clear penalties on the project team for failing to deliver them. Instead, BREEAM depends on a promise of recognition, a sense of professionalism and individual commitments to green building.

This finding raises two more general points about neoliberal techniques of governance. The first concerns the importance of their insertion in a broader system of rewards and sanctions. The neo-foucauldian literature focuses on the role of governance techniques in the constitution of subjects and the effectiveness of normative control. In doing so, it often overlooks the extent to which this effect depends on the insertion of those mechanisms in a network of interrelated practices and mechanisms. To give but one example, in the case of school exams in the United Kingdom, results are used not only to determine university entrance but also to inform parental choice (through the publication of league tables), to evaluate teacher performance (with consequences for promotions and redundancies), and to evaluate school performance (with consequences on government funding and headships). In the case of BREEAM, the final metric has yet to gain value in a network of interlocking practices, be it the property market or other systems of recognition and reward. This observation, in turn, has implications for an analysis of the relation of science and the neoliberal state. It suggests that the weaker the embeddedness of neoliberal techniques, the greater their reliance on normative and cognitive forms of legitimation and the greater the importance of (public acceptance of) scientific and technical claims for specific idea-practice couplets.

Finally, turning to the category of "identity," the case of BREEAM draws attention to the importance of weakly institutionalized standards in the exercise of neoliberal governance. At the moment, much of neo-foucauldian literature focuses on strongly embedded techniques which not only direct individual behavior but also constitute subjects (e.g., MacLullich, 2003). In this literature, subjects are presented as either internalizing dominant norms and definitions of the situation or resisting them. The case of BREEAM highlights a different relation between ideas, instruments and practices, one which rests on conscious choices rather than taken-for-granted rules.

In the case study presented above, the greater the personal engagement of the construction professional with issues of sustainable construction and the closer the proposed practice to their own area of expertise, the less likely they were to accept BREEAM practices as measures of "green building." While this effect could lead to a rejection of BREEAM practices by sustainability minded specialist engineers and construction professionals at the top of their game, it did not. Instead, sustainability minded professionals were more likely to use BREEAM to promote sustainability (along with other tools) and more likely to use it in design and construction decisions. In contrast, less committed and informed professionals were happy to accept BREEAM as scientifically committed and informed professionals, but far less likely to use it when making design and construction decisions. Whereas Foucauldians focus on the internalization of instrument related norms, the case of BREEAM points to a different type of neoliberal control, involving the critical adoption of instruments to further already existing commitments (be it to education or sustainability).

This relatively weaker effect can be linked to the relatively weak embeddedness of the instrument. Property values do not depend on BREEAM scores; jobs, and promotions are not linked to assessment outcomes; and other criteria of "best practice" are not currently overridden by BREEAM related criteria. Stated differently, for the moment, the "stakes" are low. Taken together, this combination of critical engagement by sustainability minded professionals and scientific or technical legitimation by less well-informed professionals (and the public) would seem to support the gradual incorporation of BREEAM credits into taken-for-granted rules regarding standard and best practice.

CONCLUSION

This paper began with the suggestion that BREEAM could be used as a case study to examine the role of science and scientific authority in the neoliberal use of voluntary self-regulation for the coordination of industry activity. It also suggested that BREEAM and sustainable construction more generally provided an example of the role of such instruments in the introduction of a new institutional logic. This suggestion, in turn, shifted SNI theorizing from a focus on ideas and the structural conditions for their introduction by institutional entrepreneurs to more diffuse processes, whereby the link between ideas and practices (rules) was specified on the ground through a process of experimentation.

In closing, it is helpful to revisit these claims. BREEAM as a case poses a number of challenges to theorizing both the relation between science and the neoliberal state and SNI accounts of institutional work. For NPSS it suggests that, if we care about the political effect of science, then it's important to extend research beyond the study of policy formulation to policy implementation and the more diffuse effects of science and scientific authority on everyday life. This move, however, is fraught with difficulties, starting with the specification of what is science outside of the relatively well delineated fields of science and politics. The discussion above deals with this issue by equating science and technical expertise. This rests on subjects' own conflation of the two. If one looks at how construction professionals reason, they rarely talk about science. Instead it is present by default, as a legitimate basis for particular practices. In industry, science offers a cultural signal for substantive or legitimate bases of action in opposition to purely bureaucratic or commercial considerations.

A second issue to emerge from the case study involves the instances and conditions under which scientific legitimation (of this diffuse kind) matters for neoliberal governance. The answer, suggested by the case study, is that its importance varies with the embeddedness of neoliberal techniques in broader incentive structures. The more embedded the technique – or rather the final metric which it produces and which circulates in other fields – the less important scientific or substantive legitimation. The more weakly embedded the technique, the greater its importance.

When it comes to sustainable construction, the link of specifically "sustainable" practices and market mechanisms remains weak. Carbon trading and taxes have some effect on construction clients, but not on construction firms, engineering consultancies or architectural firms whose own carbon footprint is below the United Kingdom and EU thresholds. Similarly, social investment funds are important drivers of corporate sustainability for large multinational firms, but the link between project-level practice and reporting indices is tenuous at best (most indicators focusing on firm-level rather than project-level activity). Finally, as the case of BREEAM illustrates, assessment methods and certification schemes have yet to play a significant role in the property market. Turning to dynamics within the workplace, BREEAM and similar techniques are only weakly aligned with workplace incentive structures. The result of this weak embeddedness is to render cognitive legitimation more important for the way in which professionals engage with the instrument.

If BREEAM is so weakly embedded does it matter? or rather, how does it matter? The answer suggested here is that the primary effect of neoliberal instruments like BREEAM is on the inflection of everyday practices, and more specifically on informal standards. This claim calls for a broadening of both the neo-foucauldian research agenda, which tend to focus on the effect of such instruments on subjects rather than practices, and SNI, which tends to privilege the role of actors in institutional change.

So what does this expanded version of SNI contribute to an understanding of institutional change and more specifically of institutional work around sustainable construction? If we accept an SNI specification of the phenomenon – that what we are observing and participating in is an attempt by societal groups and the state to introduce a new institutional logic into the construction and property sectors – then the case of BREEAM highlights the role of instruments in that institutional work. This suggestion, in turn, calls for an expansion of the research agenda to examine the way in which instruments mediate the association of ideas (in this case green building) with practices. While much of the research on assessment methods analyses them from an engineering and management perspective, focusing on either their formal features or on the technical characteristics of the buildings which they certify (but only partially shape), an SNI/neo-foucauldian approach draws attention to the way in which such instruments figure in the specification of policy at the field level and everyday activity at the firm level. At the field level, it highlights the way in which principled discussions about sustainable development are being narrowed in the selection and development of mandated policy mechanisms.

At the firm (and project) level, it reveals the myriad of conversations and challenges which the introduction of such instruments has elicited as well as the selective incorporation of certain criteria into increasingly taken-for-granted standards of "best practice." It also calls for much greater attention to the way in which that emerging consensus is feeding back into the ongoing specification of the instrument by the BRE and changing state support for its use.

NOTE

1. The breadth and coherence of this approach is clearly documented in the recent five-volume compilation of Greenwood, Sahlin-Anderson, Suddaby, and Oliver (2012). For a recent overview of the field, see Forgues, Greenwood, Marti, Monin, and Walgenback (2012).

REFERENCES

Amis, J., Slack, T., & Hinings, C. R. (2004). The pace, sequence and linearity of radical change. *Academy of Management Journal*, *47*, 15–39.

Atkinson, G. (2009). *Sustainable development policy: A brief review of the literature on current practice*. Final Report to the Government Economic Service Group on Sustainable Development.

Ball, S. J. (2013). *Foucault, power and education*. Oxon: Routledge.

Battilana, J., & Casciaro, T. (2012). Change agents, networks and institutions: A contingency theory of organizational change. *Academy of Management Journal*, *55*, 381–398.

Battilana, J., Leca, B., & Boxenbaum, E. (2009). How actors change institutions: Towards a theory of institutional entrepreneurship. *Academy of Management Annals*, *3*, 65–107.

Boiral, O. (2012). ISO certificates as organizational degrees? Beyond the rational myths of the certification process. *Organization Studies*, *33*, 633–654.

Botzem, S., & Dobusch, L. (2012). Standardization cycles: a process perspective on the formation and diffusion of transnational standards. *Organization Studies*, *33*(5–6), 737–762.

Bourdieu, P. (1989). Social space and symbolic power. *Sociological Theory*, *7*, 14–25.

Boxenbaum, E., Georg, S., Rieijonen, S., & Garza De Linde, G. L. (2013). *How standards enable the emergence of sustainable construction as a new organizational field*. Orlando, FL: Academy of Management.

BRE Global. (2013a). BREEAM in numbers [Online]. Retrieved from http://www.breeam.org/page.jsp?id=559. Accessed on May 25, 2013.

BRE Global. (2013b). What is BREEAM? [Online]. Retrieved from http://www.breeam.org/about.jsp?id=66. Accessed on May 25, 2013.

Brunsson, N., Rasche, A., & Seidl, D. (2012). The dynamics of standardization: Three perspectives on standards in organization studies. *Organization Studies*, *33*, 613–632.

Callon, M. (2009). Civilizing markets: Carbon trading between in vitro and in vivo experiments. *Accounting, Organizations and Society*, *34*, 535–548.

Campbell, J. L., & Pedersen, O. K. (2001). *The rise of neoliberalism and institutional analysis*. Princeton, NJ: Princeton University Press.

Centeno, M. A., & Cohen, J. N. (2012). The arc of neoliberalism. *Annual Review of Sociology*, *38*, 317–340.

Cole, R. J. (2005). Building environmental assessment methods: Redefining intentions and roles. *Building Research & Information*, *35*, 455–467.

Cole, R. J. (2006). Shared markets: Coexisting building environmental assessment methods. *Building Research & Information*, *34*, 357–371.

Cotoi, C. (2011). Neoliberalism: A Foucauldian perspective. *International Review of Social Research*, *1*, 109–124.

Courtney, R. (1997). Building Research Establishment – Past, present and future. *Building Research & Information*, *25*, 285–291.

Czarniawska, B. (2008). How to misuse institutions and get away with it: Some reflections on the institutional theory(ies). In R. Greenwood, O. Ortiz, & R. Suddaby (Eds.), *The Sage handbook of organizational institutionalism*. London: Sage.

Czarniawska, B. (2009). Commentary: STS meets MOS. *Organization*, *16*, 155–160.

D'Aunno, T., Succi, M., & Alexander, J. (2000). The role of institutional and market forces in divergent organizational change. *Administrative Science Quarterly*, *45*, 679–703.

Dean, M. (1999). *Governmentality: Power and rule in modern society*. London: Sage.

Dean, M. (2007). *Governing societies*. Maidenhead, UK: Open University Press.

Deephouse, D. L., & Suchman, M. C. (2008). Legitimacy in organizational institutionalism. In R. Greenwood, C. Oliver, K. Sahlin, & R. Suddaby (Eds.), *The handbook of organizational institutionalism*. Thousand Oaks, CA: Sage.

DEFRA. (2005). *Securing the future: Delivering UK sustainable development strategy*. London: HMSO.

DiMaggio, P. J. (1991). Constructing an organizational field as a professional project: US art museums, 1920–1940. In W. W. Powell & P. J. DiMaggio (Eds.), *The new institutionalism in organizational analysis*. Chicago, IL: University of Chicago Press.

Du Plessis, C., & Cole, R. J. (2011). Motivating change: Shifting the paradigm. *Building Research and Information*, *39*, 436–449.

Egan, J. (1998). Rethinking construction. Report of the Construction Task Force to the Deputy Prime Minister, John Prescott, on the scope for improving the quality and efficiency of UK construction, London.

Epstein, S. (1998). *AIDS, activism and the politics of knowledge*. Los Angeles, CA: University of California Press.

Ezrahi, Y. (1990). *The descent of Icarus: Science and the transformation of contemporary democracy*. Cambridge, MA: Harvard University Press.

Fiss, P. C., & Zajac, E. J. (2006). The symbolic management of strategic change: Sensegiving via framing and decoupling. *Academy of Management Journal*, *49*, 1173–1193.

Forgues, B., Greenwood, R., Marti, I., Monin, P., & Walgenback, P. (2012). New institutionalism: Roots and buds. *M@na@gement*, *15*, 459–467.

Foucault, M. (1977). *Discipline and punish: The birth of the prison*. London: Penguin.

Foucault, M. (1991[1978]). Governmentality. In G. Burchell, C. Gordon, & P. Miler (Eds.), *The Foucault Effect: Studies in governmentality*. London: Harvester Wheatsheaf.

Foucault, M. (2008). *The birth of biopolitics: Lectures at the College de France*. London: Palgrave Macmillan.

Fourcade, M. (2009). *Economists and societies: Discipline and profession in the United States, Britain and France, 1890s to 1990s*. Princeton, NJ: Princeton University Press.

Freiburger, N., Biggart, N. W., & Beamish, T. D. (2012). Hidden in plain sight: Towards an understanding of the liberal state/free market nexus. *Sustainable construction workshop*, Ecole des Mines, Paris.

Friedland, R. (2012). Book review: The institutional logics perspective. *M@na@gement*, *15*, 582–595.

Fuerst, F., & McAllister, P. (2011). The impact of Energy Performance Certificates on the rental and capital values of commercial property assets. *Energy Policy*, *39*, 6608–6614.

Gamble, A. (1988). *The free economy and the strong state*. Durham, NC: Duke University Press.

Gane, N. (2012). The governmentalities of neoliberalism: Panopticism, post-panopticism and beyond. *Sociological Review*, *60*, 611–634.

Gibson, R., Hassan, S., Holtz, S., Tansey, J., & Whitelaw, G. (2005). *Sustainability assessment: Criteria and processes*. London: Earthscan.

Gilad, S. (2012). *How firms translate regulatory messages*. Centre for Analysis of Risk and Regulation, London School of Economics.

Gouldson, A., & Bebbington, J. (2007). Corporations and the governance of environmental risk. *Environment and Planning C: Government and Policy*, *25*, 4–20.

Green, S. D. (2011). *Making sense of construction improvement*. Chichester: Wiley-Blackwell.

Greenwood, R., & Hinings, C. R. (1996). Understanding radical organizational change: Bringing together the old and the new institutionalism. *Academy of Management Review*, *21*, 1022–1054.

Greenwood, R., Micelotta, E. R., Raynard, M., Kodeih, F., & Lounsbury, M. (2011). Institutional complexity and organizational responses. *The Academy of Management Annals*, *5*, 317–371.

Greenwood, R., Sahlin-Anderson, K., Suddaby, R., & Oliver, C. (Eds.). (2012). *Institutional theory in organization studies*. London: Sage.

Greer, J., & Bruno, K. (1996). *Greenwash: The reality behind corporate environmentalism*. Penang, Malaysia: Third World Networking.

Hallett, T., & Ventresca, M. J. (2006). Inhabited institutions: Social interactions and organizational forms in Gouldner's patterns of industrial bureaucracy. *Theory and Society*, *35*, 213–236.

Hamilton, A., & Gioia, D. A. (2009). Fostering sustainability-focused organizational identities. In L. M. Roberts & J. E. Dutton (Eds.), *Exploring positive identities and organizations: Building a theoretical and research foundation*. New York, NY: Taylor and Francis Group.

Hardy, C., & Maguire, S. (2008). Institutional entrepreneurship. In R. Greenwood, O. Ortiz, & R. Suddaby (Eds.), *The Sage handbook of organizational institutionalism*. London: Sage.

Hasselbladh, H., & Kallinikos, J. (2000). The project of rationalization: A critique and reappraisal of neo-institutionalism in organization studies. *Organization Studies*, *21*, 697–720.

Hess, D. J. (2007). *Alternative pathways in science and industry: Activism, innovation and the environment in an era of globalization*. Cambridge, MA: MIT Press.

Hess, D. J. (2011). To tell the truth: On scientific counterpublics. *Public Understandings of Science*, *20*, 627–641.

Hoffman, A. J. (1999). Institutional evolution and change: Environmentalism and the U.S. chemical industry. *The Academy of Management Journal, 42*, 351–371.
Hoffman, A. J. (2001). Linking organizational and field-level analyses: The diffusion of corporate environmental practice. *Organization & Environment, 14*, 133–156.
Hoffman, A. J., & Georg, S. (2012). A history of research on business and the natural environment: Conversations from the field. In A. J. Hoffman & S. Georg (Eds.), *Business and the natural environment*. Oxford: Routledge.
Hoffman, A. J., & Henn, R. (2008). Overcoming the social and psychological barriers to green building. *Organization Environment, 21*, 390–419.
Kajikawa, Y., Inoue, T., & Goh, T. N. (2011). Analysis of building environment assessment frameworks and their implications for sustainability indicators. *Sustainability Science, 6*, 233–246.
Kerwer, D. (2005). Rules that many use: Standards and global regulation. *Governance, 18*, 611–632.
Kinchy, A. J. (2012). *Seeds, science and struggle: The global politics of transgenic crops*. Cambridge, MA: MIT Press.
Kinchy, A. J., Kleinman, D. L., & Autry, R. (2008). Against free markets, against science? Regulating the socio-economic effects of biotechnology. *Rural Sociology, 73*, 147–180.
Kleinman, D. L., Kinchy, A. J., & Autry, R. (2009). Local variation or global convergence in agricultural biotechnology policy? A comparative analysis. *Science and Public Policy, 36*, 361–371.
Kleinman, D. L., Kinchy, A. J., & Handelsman, J. (2005). *Controversies in science and technology, volume one: From maize to menopause*. Madison, WI: The University of Wisconsin Press.
Kraatz, M. S., & Block, E. S. (2008). Organizational implications of institutional pluralism. In R. Greenwood, O. Ortiz, & R. Suddaby (Eds.), *The Sage handbook of organizational institutionalism*. London: Sage.
Latham, M. (1994). *Constructing the team: Final report of the government/industry review of procurement and contractual arrangements in the UK construction industry*. London: HMSO.
Lawrence, T. B., Suddaby, R., & Leca, B. (2009). *Institutional work: Actors and agency in institutional studies of organizations*. Cambridge: Cambridge University Press.
Lemke, T. (2001). "The birth of bio-politics": Michel Foucault's lecture at the Collège de France on neo-liberal governmentality. *Economy and Society, 30*, 190–207.
Linnenluecke, M. K., & Griffiths, A. (2013). Firms and sustainability: Mapping the intellectual origins and structure of the corporate sustainability field. *Global Environmental Change, 23*, 382–391.
Lohman, L. (2009). Toward a different debate in environmental accounting: The cases of carbon and cost-benefit. *Accounting, Organizations and Society, 34*, 499–534.
Lounsbury, M. (2007). A tale of two cities: Competing logics and practice variation in the professionalizing of mutual funds. *Academy of Management Journal, 50*, 289–307.
Lounsbury, M., & Crumley, E. T. (2007). New practice creation: An institutional perspective on innovation. *Organization Studies, 28*, 993–1012.
Lützkendorf, T., & Lorenz, D. (2011). Capturing sustainability-related information for property valuation. *Building Research & Information, 39*, 26–273.
MacKenzie, D. (2009). Making things the same: Gases, emission rights and the politics of carbon markets. *Accounting, Organizations and Society, 34*, 440–455.

MacLullich, K. K. (2003). The emperor's "new" clothes? New audit regimes: Insights from Foucault's technologies of the self. *Critical Perspectives on Accounting*, *14*, 791–811.

Meyer, J., Boli, J., & Thomas, G. (1987). Ontology and rationalization in western cultural account. In G. M. Thomas, J. W. Meyer, & F. O. Ramirez (Eds.), *Institutional structure: Constituting state, society and the individual*. Newbury Park, CA: Sage.

Meyer, J. W., & Rowan, B. (1977). Institutionalized organizations: Formal structure as myth and ceremony. *American Journal of Sociology*, *83*, 340–363.

Miller, P., & Rose, N. (1990). Governing economic life. *Economy and Society*, *19*, 1–31.

Moore, K., Kleinman, D. L., Hess, D., & Frickel, S. (2011). Science and neoliberal globalization: A political sociological approach. *Theoretical Sociology*, *40*, 505–532.

Moran, M. (2003). *The British regulatory state: High modernism and hyper innovation*. Oxford: Oxford University Press.

Newman, J. (Ed.). (2005). *Remaking governance: Peoples, politics and the public sphere*. Bristol: The Policy Press.

OGC. (2008). Sustainable procurement and operations on the government estate: Government delivery plan. HM Government.

Oliver, C. (1991). Strategic responses to institutional processes. *Academy of Management Review*, *16*, 145–179.

Pache, A.-C., & Santos, F. (2010). When worlds collide: The internal dynamics of organizational responses to conflicting institutional demands. *Academy of Management Review*, *35*, 455–476.

Pestre, D. (2008). Challenges for the democratic management of technoscience: Governance, participation and the political today. *Science as Culture*, *17*, 101–119.

Phillips, N., Lawrence, T. B., & Hardy, C. (2004). Discourse and institution. *Academy of Management Review*, *29*, 635–653.

Porter, T. (1995). *Trust in numbers: The pursuit of objectivity in science and public life*. Princeton, NJ: Princeton University Press.

Potoski, M., & Prakash, A. (2004). The regulation dilemma: Cooperation and conflict in environmental governance. *Public Administration Review*, *64*, 152–163.

Potoski, M., & Prakesh, A. (2005). Green clubs and voluntary governance: ISO 14001 and firms' regulatory compliance. *American Journal of Political Science*, *49*, 235–248.

Power, M. (1999). *The audit society*. Oxford: Oxford University Press.

Prozorov, S. (2007). *Foucault, freedom and sovereignty*. Aldershot: Ashgate.

Rose, N., & Miller, P. (1992). Political power beyond the state: Problematics of government. *British Journal of Sociology*, *43*, 172–205.

Sandholtz, K. W. (2012). Making standards stick: A theory of coupled vs. decoupled compliance. *Organization Studies*, *33*, 655–679.

Schweber, L. (2013). The effect of BREEAM on clients and construction professionals. *Building Research & Information*, *41*, 129–145.

Schweber, L. (2014). Putting theory to work: The use of theory in construction research. *8th CIDB postgraduate conference*, University of the Witwatersrand, Johannesburg, South Africa.

Schweber, L., & Haroglu, H. (2014). On the fit between assessment and design processes: A comparison of eight BREEAM case studies. *Building Research & Information*, *42*(3), 300–317.

Scott, W. R. (1995). *Institutions and organizations*. Thousand Oaks, CA: Sage.

Short, J. L., & Toffel, M. W. (2010). Making self-regulation more than merely symbolic: The critical role of the legal environment. *Administrative Science Quarterly*, *55*, 361–396.

Slager, R., Gond, J.-P., & Moon, J. (2012). Standardization as institutional work: The regulatory power of a responsible investment standard. *Organization Studies*, *33*, 763–790.

Suchman, M. C. (1995). Managing legitimacy: Strategic and institutional approaches. *Academy of Management Review*, *20*, 571–610.

Thornton, P. H., & Ocasio, W. (1999). Institutional logics and the historical contingency of power in organizations: Executive succession in the higher education publishing industry, 1958–1990. *American Journal of Sociology*, *105*, 801–843.

Thornton, P. H., Ocasio, W., & Lounsbury, M. (2012). *The institutional logics perspective: A new approach to culture, structure and process*. Oxford: Oxford University Press.

Timmermans, S., & Epstein, S. (2010). A world of standards but not a standard world: Toward a sociology of standards and standardization. *Annual Review of Sociology*, *36*, 69–89.

Tryggestad, K., & Georg, S. (2011). How objects shape logics in construction. *Culture and Organization*, *17*, 181–197.

U.N. World Commission on Environment and Development. (1987). Our common future. The Bruntland Report. Oxford University Press, Oxford.

Weik, E. (2012). Introducing "The creativity of action" into institutionalist theory. *M@na@gement*, *15*, 563–581.

Whittington, R. (2006). Completing the practice turn in strategy research. *Organization Studies*, *27*, 613–634.

Wooten, M., & Hoffman, A. J. (2008). Organizational fields: Past, present and future. In R. Greenwood, O. Ortiz, & R. Suddaby (Eds.), *The Sage handbook of organizational institutionalism*. London: Sage.

Zilber, T. B. (2002). Institutionalization as an interplay between actions, meanings and actors: The case of a rape crisis center in Israel. *Academy of Management Journal*, *45*, 234–254.

FIELD THEORIES AND THE MOVE TOWARD THE MARKET IN US ACADEMIC SCIENCE

Elizabeth Popp Berman

ABSTRACT

Field theory is one of the most visible approaches in the new political sociology of science, and Fligstein & McAdam's (F&M) Theory of Fields *is the most visible recent attempt to further it. This paper evaluates F&M's theory of field transformation by comparing it with Berman's (2012a) field-based explanation of the changes in the field of US academic science. While F&M's general framework is quite useful, their explanation, which focuses on struggles between incumbents and challengers over whose conception of the field should dominate, does not map neatly onto the changes in academic science, which saw no such field-level struggles. This suggests that tools are also needed for explaining new settlements that do not result from intentional efforts to establish them. In particular, the case of US academic science shows that local innovations with practices based on alternative conceptions of the field can lead to field-level change. Attention to the interaction between local*

Fields of Knowledge: Science, Politics and Publics in the Neoliberal Age
Political Power and Social Theory, Volume 27, 193–221
Copyright © 2014 by Emerald Group Publishing Limited
All rights of reproduction in any form reserved
ISSN: 0198-8719/doi:10.1108/S0198-871920140000027015

practice innovations and larger environments provides insights into how change ripples across fields, as well as the ongoing contention and dynamism within even relatively stable fields.

Among the variety of approaches taken by the new political sociology of science (Frickel & Moore, 2006), field theory is prominent (Albert & Kleinman, 2011; Berman, 2012a; Hess, 2012). But different versions of field theory have been developed, and additional variations continue to be proposed. In the sociology of science, both Bourdieusian field theory (Albert & Kleinman, 2011; Camic, 2011; see also Albert and McGuire, 2014) and organizational field theory (Berman, 2012a, 2012b; Owen-Smith, 2011; Swan, Bresnen, Robertson, Newell, & Dopson, 2010) have been used.

Recently, Fligstein and McAdam (F&M) (2011, 2012) have made a high-profile attempt to integrate and move beyond existing versions of field theory. Their framework, which explicitly draws on both Bourdieu and the organizational tradition, but also uses insights from economic sociology, historical institutionalism, and the social movements literature, has ambitions to build a common foundation for thinking about how social action leads to the creation and transformation of mesolevel social orders: that is, fields.

This paper explores the potential of that framework by applying it to an empirical case familiar to sociologists of science, and which I have elsewhere (Berman, 2012a, 2012b) used organizational field theory to explain: the changes in US academic science that have taken place since the 1970s, which have variously been characterized as "mode 2" (Gibbons et al., 1994), a "triple helix" (Leydesdorff & Etzkowitz, 1996), "academic capitalism" (Slaughter & Leslie, 1997), "asymmetric convergence" (Kleinman & Vallas, 2001), and "neoliberal" (Lave, Mirowski, & Randalls, 2010).

Renarrating this story using F&M's framework, and then explicitly comparing it with my own, shows the strengths and limitations of each of these approaches to field theory, as well as highlighting ways that the two might be fruitfully extended or combined. In particular, F&M bring two major advantages over other variants of field theory: their conceptualization of fields as nested and overlapping, and their focus on the relationship *between* fields as a main source of change. They also bring a much-needed attention to power that is central to Bourdieu but tends to be played down in organizational field theory. But their framework, at least as presented in

A Theory of Fields, also has two limitations that the empirical case of US academic science highlights: an overemphasis on field-level strategic action and a focus on transitions from one settled state to another.

A conversation between the two approaches points to places where the strengths of each can be combined. In particular, the organizational approach to fields has increasingly developed tools for explaining change from the bottom up and change initiated outside the field (Berman, 2012b; Greenwood, Díaz, Li, & Lorente, 2010; Purdy & Gray, 2009; Smets, Morris, & Greenwood, 2012). These insights can be used in a way that is compatible with F&M's framework, yet goes beyond its present form. At the same time, F&M draw needed attention to the fact that bottom-up and externally driven field change may actually reflect relationships between nested, overlapping, or adjacent fields. Thus considering multiple versions of field theory can help us think more deeply about nature of the link between fields, a promising area for further research.

The paper develops this argument in five parts. First, it provides a short empirical introduction to the changes in US academic science over the past few decades. Second, it reviews major field-theoretic approaches and how they suggest we should think about those changes. Third, it retells the story of the transformation of US academic science using F&M's framework. Then it more briefly examines that story through my own lens, with particular attention to the points of difference between my version and the one suggested by *A Theory of Fields*. Finally, it explicitly compares the two approaches: their strengths, weaknesses, and points of overlap.

THE CHANGING FIELD OF US ACADEMIC SCIENCE: ACTORS, POWER, BELIEFS, AND PRACTICES

Any empirical description of a field must identify the major actors within it, how power and resources are distributed among them, what they understand the purpose of the field to be, and what kinds of action they perceive to be legitimate. Over the last thirty-five years, the field of US academic science has seen the perceived purpose of the field change substantially, as have accepted practices within it. The distribution of power and resources has evolved more moderately, and there has been little change in the composition of the major actors within the field. The rest of this section, based on a summary of previous research (Berman, 2012a), sketches changes and continuities in each of these areas.

Adoption of New Beliefs

From 1978 to the present, a new set of assumptions about how science gets into practical use, and the role played by universities in that process, has been widely adopted within US academic science. These include the following:

(1) Scientific inventions often originate in academia but should be moved into the private sector as quickly as possible for the most efficient development of applications.
(2) A system of incentives, such as patent rights and ownership stakes in startup companies, is needed to encourage both academic scientists and industry to work toward these ends.
(3) As the source of many scientific inventions, academic scientists have a unique role to play in the process of technology transfer.
(4) Universities have a public responsibility to encourage and facilitate this process.
(5) Doing so is important not only because it will get science rapidly into use, but also because it will maximize science's contribution to economic growth and competitiveness.

This way of thinking about the purpose of academic science has partially displaced other views. For example, one alternative belief that has become less common is that the purpose of academic science is simply to produce knowledge. From this perspective, while society supports scientific research partly because of its effects on the world, the best way to ensure those effects is to insulate science and allow it to develop on its own. Identifying and facilitating practical applications should not be a priority of academic scientists. Another alternative view is that society supports academic research to solve specific problems identified by government, like defense needs or health issues. Both of these understandings can still be seen within universities but have lost influence in comparison with the perspective that emphasizes technology transfer and economic impact.

Adoption of New Practices

This new sense of the field's purpose is reflected in a variety of practices focused on realizing the economic value of academic science. These were once uncommon and of questionable legitimacy but have spread and gained acceptance over the last several decades. They include the patenting and licensing of scientific inventions and the creation of university-sponsored spinoff companies, faculty entrepreneurship, the creation of research parks and startup incubators to encourage university–industry relationships, the establishment of a greater variety of university–industry partnerships, and the development of incentives for faculty who pursue activities aimed at technology transfer, such as applying for patents. More generally, a variety of routine scientific activities have been tweaked to play up their potential for commercial applications or their impact on economic development.

Evolving Sources of Power

While beliefs and practices have evolved significantly in US academic science, power within the field still rests on the same sources: scientific authority and economic resources. Scientific authority is conferred upon individuals by professional communities within science. It is produced through publication in high-status journals, through association with high-status peers and institutions, and through professional acknowledgment in the form of citations and awards that recognize the production of scientific knowledge valued by the community. The list of universities, journals, and professional associations most important for conferring scientific authority has much overlap with the list from fifty years ago.

There are also significant continuities in the providers of economic resources. In particular, government, led by the National Institutes of Health (NIH), the National Science Foundation (NSF), and the Department of Defense (DOD), still pays for about two-thirds of academic R&D, down from about three-quarters in 1978. The modest decrease has been made up not by industry funding, which has increased from 4% to 6% of the total, but by academic institutions themselves (which provide 20% of all support today, vs. 14% in 1978) (National Science Board, 2012, Appendix Table 5-2).

There have been moderate, but important, changes in how both scientific and economic capital are acquired in the field. First, while it is more or less the same group of institutions that confer scientific authority, there has been some evolution in what kinds of contributions those institutions value. In particular, academic scientists and their institutions see applied, commercially relevant research as having greater worth now than they did several decades ago, when many saw doing such research as selling out (Boyer, 1994).

Second, the sources of economic capital have also changed in some ways. NIH has become more important to the resource environment, providing 39% of all US academic R&D funding today, versus 29% in 1978. Federal agencies other than NIH, NSF, and DOD have become less important: they once supported 19% of academic research but now fund only 6% of the total (National Science Board, 2012, Appendix Tables 5-2, 5-3). But more critical than the mix of agencies that support academic science is the evolution of their priorities. NSF, for example, was once exclusively oriented toward support for fundamental research, but today focuses on "programs that help drive future economic growth, global competitiveness, and the creation of high-wage jobs for American workers" (National Science Foundation [NSF], 2013, p. 1). Thus while the sources of support have evolved only modestly, the broader shift toward a model of science focused on technology transfer and economic development means that research that can be framed in such terms has a funding advantage, even when support is not explicitly targeting such projects.

But Similar Players and Rules

While new beliefs and practices are quite visible in academic science, and the distribution of resources has changed in some ways, the major actors, and types of actors, in the field are largely the same. A few additional universities – for example, Georgia Tech and the University of California, San Diego – have entered the top tier of scientific institutions. Some new professional organizations, like the Association of University Technology Managers, have become visible. The congressional committees and White House offices that supervise science policy have evolved. But for the most part, the actors important to the field today would be quite recognizable to anyone who was involved with it in the 1970s.

Similarly, while there have been important laws passed, like the Bayh–Dole Act, that have changed specific rules governing the field, there has been no dramatic rethinking of what agencies will provide support for scientific research, or of the role of peer review in making those decisions. Thus if one wants to explain changes in the field of academic science, the most important ones to explain are the evolution of beliefs about the field's purpose, how it works, and how it relates to the larger world, along with the emergence of specific practices tied to those beliefs.

SURVEYING FIELD THEORIES OF CHANGE

The evolution that has taken place in US academic science is important, well-examined, and lends itself to a field-theoretic approach. But different versions of field theory, while overlapping, suggest different ways of thinking about how fields change.

Organizational theory provides one framework for thinking about fields (DiMaggio & Powell, 1983; Greenwood, Suddaby, & Hinings, 2002; Leblebici, Salancik, Copay, & King, 1991), and I have used this framework elsewhere to understand the changes in US academic science (Berman, 2012a, 2012b). Bourdieusian field theory (Bourdieu, 1995, 1988; Bourdieu & Wacquant, 1992) provides an alternative, and sociologists of science have also drawn on Bourdieu (Albert & Kleinman, 2011; Camic, 2011; Cooper, 2009), though not specifically to explain *why* fields change. F&M's recent work (2011, 2012) attempts to build on both the organizational tradition and Bourdieu while going beyond them by incorporating insights from recent work in social movements, historical institutionalism, and economic sociology.

For Bourdieu, fields are primarily spaces of struggle for dominance, with their own stakes, rules, and forms of capital. In the field of science, the game is about acquiring scientific capital and getting to determine what counts as "good science" (Bourdieu, 1981 [1975], 2004). Fields have varying degrees of autonomy; that is, they differ in the extent to which internal actors, versus external actors with economic and political power, get to define the rules of the game and the form of capital that has value within it.

Within Bourdieu's framework, field-level change can be understood in two ways. First, fields change as the result of struggles between different groups of actors within the field. Each wants to see its own attributes defined as the legitimate form of capital. And each will work, in alliance with groups in other, connected fields, to change the evaluation criteria to favor its own form of capital. Second, field change can result from hysteresis – a mismatch between a relatively durable habitus and changing environmental conditions that can trigger unpredictable reactions (Bourdieu, 2000). There has recently been a move toward applying Bourdieu to the sociology of science (see particularly Albert & Kleinman, 2011). But no one has yet used Bourdieu specifically to explain the changes in US academic science in recent decades (though see Albert & McGuire, 2014; Lave, 2012; McGuire, 2011 for related work), and this paper does not try to fill that gap.

Like Bourdieu, organization theory sees fields as structured, relational social spaces in which actors play a common game with agreed-upon rules. But while organizational field theory was inspired in part by Bourdieu's work (Mohr, 2005), there are important differences between the two (see Dobbin, 2008; Emirbayer & Johnson, 2008; Vaughan, 2008 on the intersection of Bourdieu and organization theory). Compared to Bourdieusian approaches, organizational theories unsurprisingly pay more attention to the role of organizations in structuring fields. Less obviously, though, they also focus much less on power (Emirbayer & Johnson, 2008). Instead, they have particularly emphasized institutional entrepreneurship (Hardy & Maguire, 2008) and social movement dynamics (Schneiberg & Lounsbury, 2008) as sources of change.

More recently, however, organizational field theory has shifted its focus toward types of change that are more evolutionary and incremental, and less strategic and intentional (Purdy & Gray, 2009). Much of this work focuses on the emergence and diffusion of new practices (Lounsbury & Crumley, 2007; Smets et al., 2012) and on how competing and hybrid institutional logics create possibilities for change (Greenwood et al., 2010; Greenwood, Raynard, Kodeih, Micelotta, & Lounsbury, 2011; Thornton, Ocasio, & Lounsbury, 2012).

My own explanation of the changes in US academic science is grounded in, and illustrates, this recent scholarship. Following Friedland and Alford (1991), I begin with the assumption that major institutional logics are broadly available in modern society. I further assume that in any field some people are always experimenting with new practices and that such practices instantiate one institutional logic or another. But while multiple logics are always available, some will be stronger than others in a particular field: they will be more visible, and practices based on them will be easier to reproduce. When the field's environment changes, it may become easier for practices based on alternative logics to take hold and spread. If such practices become common enough, the alternative logic may gain strength in the field more broadly (Berman, 2012b).

This, I argue, is what happened within the field of US academic science. Before the 1970s, one could find individuals experimenting with practices based on market logic – practices that saw the value of science in its potential to create economic value – even at a time when other logics that valued science for its own sake or for its service to the state were stronger. In the late 1970s, though, the environment beyond the field – in particular, the preferences of policymakers – made it easier for market-logic practices to take hold and created resources that helped them to do so. The result was the gradual, but profound, strengthening of market logic within the scientific field, which took place even though there was no entrepreneurial or social-movement-like project aimed at changing the field. This is consistent with other recent work in organization theory that emphasizes the interaction between individual innovation, multiple logics, and the broader environment (see Thornton et al., 2012, chapter 6 for a review).

F&M, however, want to go beyond the organizational conception of fields, critiquing it for an inability to explain change other than through the "great man" of the institutional entrepreneur, a lack of attention to fields' larger environments, and inattention to power (2012, pp. 28–29). Instead, they propose their own synthesis, focusing particularly on field emergence and field transformation, rather than on periods of relative stability. Their framework emphasizes seven concepts: (1) fields as, first and foremost, spaces of strategic action; (2) the role of incumbents, challengers, and governance units; (3) social skill and the existential functions of the social; (4) the broader field environment; (5) exogenous shocks, mobilization, and the onset of contention; (6) episodes of contention; and (7) settlement (F&M, 2012, pp. 8–9). F&M see fields as structured by rules set up to favor, and defended by, incumbents, but which challengers are always trying to change to favor themselves.

So far, this is compatible with both Bourdieu and organization theory. But in their explanation of change, F&M emphasize the relationship among fields to a greater extent than other field theories. This relationship is the most common source of change: a crisis in one field ripples out through all the others it is linked to (F&M, 2012, p. 19). The larger of these ripples create opportunities for "episodes of contention" that can lead to field-level change. In an episode of contention, challengers identify a threat to incumbents or an opportunity to advance their own interests. They mobilize their resources and use innovative forms of collective action to leverage the threat or opportunity. Incumbents seek to defend the status quo, and both challengers and incumbents work "to mobilize consensus around a particular conception of the field" (F&M, 2012, p. 22). An episode of contention ends when a new settlement is reached, often with the intervention of state actors: either a restoration of the status quo ante or the institutionalization of a new set of rules and norms favoring a new set of incumbents.[1]

Thus F&M's theory seems well-suited to explaining why change took place in the field of US academic science. It would suggest we look for exogenous shocks leading to an initial disruption of the field, perhaps coming from "fields upon which the strategic action field is dependent" (F&M, 2012, p. 99); challengers identifying this disruption as an opportunity and using their resources and social skill to promote a conception of the field more focused on commercial applications of science; incumbents defending the status quo; and eventual imposition of a new settlement, most likely enforced by the state.

In *A Theory of Fields*, F&M apply their framework to two empirical cases of field transformation: US racial politics and the US mortgage market. In each case, they break their analysis into five moments: (1) "strategic action field: before," (2) "exogenous shocks," (3) "crisis and contention," (4) "settlement and the new strategic action field," and (5) "effects on other strategic action fields" (F&M, 2012, p. 115). In the section that follows, I use this approach to tell the story of the changing field of US academic science, drawing on the research reported in Berman (2012a). The resulting narrative is incomplete but is also generative of new insights into both the empirical case and the question of how fields change more generally.

THE TRANSFORMATION OF THE STRATEGIC ACTION FIELD OF US ACADEMIC SCIENCE

Strategic Action Field: Before

The field of US academic science was in a fairly settled period from the end of World War II to the late 1960s. It was growing rapidly, of course, but that was part of the settlement: both policymakers and scientists assumed that academic science would receive double-digit budget increases every year. In this "social contract for science" (Guston, 2000), it was also assumed that academic science would serve defense purposes, and that universities would in general be responsive to the needs and preferences of the federal government (Berman, 2012a, pp. 35–37).

But one final assumption, possibly the most important of all, was also built into the settlement. Both policymakers and scientists believed that unrestricted federal support for basic research would, by virtue of serendipity, lead to practical applications (Swazey & Reeds, 1978, pp. 5–8). This assumption was ironic, given than the settlement itself grew out of the success of the Manhattan Project, the largest applied science effort of all time. Yet it was fundamental and widely shared. This was a golden period for academic science, a time of rapid expansion (National Science Board, 2010, Appendix Table 4-3) and significant autonomy.

The postwar settlement must be understood in the context of the Cold War. Defense was *the* central concern of American politics, a fact reflected in the prominence of the DOD, the Atomic Energy Commission, and NASA among university funders at the time. In 1960, nearly half of universities' federal R&D dollars came from these three agencies (NSF, 2003, Table B). Physicists, who had been central to the Manhattan Project, sat at the top of the scientific hierarchy, and high-level science advisers came disproportionately from the physics community (Kevles, 1978).

Yet the "pure science" part of this deal was less obvious. It had been wrangled by scientists in the postwar years, written into *Science – The Endless Frontier* and institutionalized in the NSF (Kleinman, 1995). While NSF, which explicitly supported basic science, provided only about 10% of universities' federal research dollars in 1960 (NSF, 2003, Table B), the principle was also upheld by DOD's Advanced Research Projects Agency, which supported "blue sky" research on the assumption that the big defense-relevant breakthroughs were unpredictable (Richard J. Barber Associates, 1975), and even, to a surprising extent, NIH (Endicott & Allen, 1953). With the nuclear bomb still recent, and in an era of expanding budgets, this view was rarely challenged.

This postwar settlement served university scientists very well. Although money flowed particularly freely in defense-related fields, budgets were generally expansionary. Departments were hiring new faculty. Scientists got to decide what kinds of knowledge were valuable. And as late as 1967, scientists thought that 15% annual growth in funding was a reasonable target for the foreseeable future (Greenberg, 1967).

Exogenous Shocks

But things were about to change. As F&M would predict, the changes that disrupted this settlement were largely external to the field of US academic science. Prominent among them was a tightening of the federal budget as the postwar economic expansion slowed and the United States became embroiled in Vietnam (Berman, 2012a, p. 36). The end of the 1960s saw not 15% annual increases in the federal science budget but slight declines (NSB, 2010, Appendix Table 4-3). After the extended period of growth, scientists perceived these cuts as quite painful (Geiger, 1993, p. 246).

Budget pressures also led to more questions about "blue-sky research." In 1966, President Lyndon Johnson suggested in a speech that he might push NIH in a more applied direction (Walsh, 1966). The Mansfield amendment declared that defense funds would be restricted to research with "a direct and apparent relationship to a specific military function or operation" (Public Law 91-121). Even NSF saw a 1968 amendment that authorized it, for the first time, to support applied research (Belanger, 1998, p. 76). All these changes called into question the settlement's assumption that if the federal government simply provided support to scientists and let them decide what to do with it, the rest would take care of itself.

Another external crisis, the escalation of student protests against the Vietnam War (Geiger, 1993, pp. 230–242), was also putting strain on the field's settlement in the late 1960s. As the gap between campuses and Washington became more visible, some members of Congress demanded that universities better control their students and threatened to cut funding, including science funding, if they did not (Boffey, 1969; Walsh, 1969). Some scientists, too, protested against the role science had come to play in the military–industrial complex (Moore, 2008). The resistance of both students and scientists helped to undermine the settlement's assumption that universities would be responsive to government, including military, needs.

Crisis and Contention

This combination of federal budget retrenchment, pressures for more applied research, and campus unrest effectively destabilized the postwar settlement that had served scientists so happily for two decades. Scientists and universities began to complain that "deep budget cuts ha[d] undermined their ability to do effective research" and that they "fear[ed] an unprecedented financial crisis" (Boffey, 1968, p. 340; 1970, p. 555). They called repeatedly for a return to the status quo ante, defending the assumptions of the old settlement:

> It may ... be an undesirable distraction to its rigor and sharpness of focus for the research worker himself to be too sensitive to the unpredictable implications of his own work ... . To place the burden of such justification [of practical utility] on individual projects would be the surest possible way of stifling the most creative, the least predictable advances in scientific understanding. (US Senate, 1967, pp. 8–9)

But scientists were slow to perceive just how much things had changed, and the field's incumbents seemed ill-positioned to develop a new vision of the field. By the mid-1970s, federal science budgets were still flat, DOD had largely stopped supporting blue-sky research, and NSF found even its strongest supporters criticizing its lack of concern with practical applications (House, Senate Split, 1977; Richard J. Barber Associates, 1975;

Smith & Karlesky, 1977). Other groups, however, were setting the stage for what would eventually resolve itself into a new settlement, and a couple of these were able to use this period of crisis to advance their own interests, as F&M suggest would happen.

One such group was working to start companies that would commercialize recombinant DNA (rDNA) technology. An easy-to-replicate method of recombining DNA was published in 1973, and within a few years a small number of academics, venture capitalists, and other investors were trying to develop applications. Public fears about the dangers of rDNA, however, as well as sharp boundaries between academia and industry were threatening their efforts (Berman, 2012a, pp. 62–71). This group wanted to see academics participate in the commercialization of rDNA research, with the ultimate goal of creating a new industry.

A second loose group wanted to promote university patenting. Motivated by a belief that patents were necessary in order to get scientific inventions into use more rapidly, some of its members were university-based and others were located in government agencies. In the mid-1970s, they formed a professional association, the Society of University Patent Administrators, which became, in F&M's language, an internal governance unit within the field of US academic science (Berman, 2012a, pp. 100–106).[2] This group hoped to change the rules and norms around patenting scientific inventions in the field of US academic science.

A third group, however, would have important effects on the field of academic science even though it was not trying to use the crisis in that field to its own advantage. In fact, it was located in another field entirely and was responding to its own field-level crisis. This group was made up of scientists and executives who managed industrial R&D operations for large corporations. The economic conditions of the 1970s, and increasing competition from Japan and Germany, were threatening the field of R&D-intensive firms. Leaders of those firms' research arms diagnosed this crisis as resulting from a failure of technological innovation and began working to draw the innovation issue to the attention of policymakers and the media (Berman, 2012a, pp. 51–54). While they were not targeting the field of academic science, their efforts would, indirectly, prove important to it.

Settlement and the New Strategic Action Field

During the late 1970s and early 1980s, the terms of a new settlement were laid out. The old order, which assumed that government would support academic science with ever-increasing amounts of money, that useful results would automatically follow, and that universities would be responsive to government preferences, came to an end.

The new order rested on a different conception of how academic science served the public interest and why government should support it. In this vision, the purpose of academic science was to create technological innovations. These innovations would lead to the creation of new companies, or result in products that would make existing companies more competitive, spawn whole new industries, and expand the economy. This process would not occur effortlessly. Universities would have to exert themselves actively to encourage commercially promising inventions and transfer them to industry. In conjunction with this change, practices like industry collaboration, entrepreneurship, and patenting became more prevalent, and those who produced science with commercial potential gained power and resources in the field (Berman, 2012a, pp. 30–32).

This new settlement was not driven by either of the two groups that were trying to use the period of crisis to advance their own interests, although they were indeed able to advance their causes – academic scientists did help establish the biotechnology industry, and the field's rules changed to encourage universities to patent their inventions. The group that proved most important to the transformation of academic science, however, the scientists and executives working for R&D-intensive companies, was the group that was not trying to change academic science at all.

To understand how that could be the case, one must realize that the conception of science that underpinned this new settlement came from in the policy field, not the academic field, and was advanced by the group representing R&D-intensive firms. These firms, which were struggling in the economic environment of the 1970s, developed a narrative that explained their own troubles and identified solutions to them. They argued that a crisis of technological innovation was causing them to lose global economic leadership. This crisis was caused by government policies, and changes in those policies – including cutting taxes, strengthening patent rights, deregulation, and loosening antitrust enforcement – would stimulate innovation and restore the position of US industry (Berman, 2012a, p. 54).

Congress went on to make all those changes, to a greater or lesser extent, in the late 1970s and early 1980s. But the leaders of these firms did not ask Congress to involve universities in solving the problem of innovation. Instead, as R&D-intensive firms convinced policymakers that innovation was both economically critical and in crisis, the policy environment became friendlier to academic practices that treated science as something of potential commercial value – as an input into the economy (Berman, 2012a, pp. 54–57). Policymakers made a variety of decisions that had incremental but pervasive effects on academic science, and activities like patenting, faculty entrepreneurship, and industry collaboration became more common. Increasingly, incumbents within the field of academic science began to use the language of innovation and the metaphor of academic science as an economic engine as ways to talk about its purpose (Berman, 2012a, pp. 154–156).

Unlike in some more radical episodes of change, the field's incumbents did not change dramatically as the new settlement stabilized. While some internal governance units, like the Association of University Technology Managers, became more visible, for the most part the new settlement saw the same agencies supporting university research, and the same organizations representing universities and scientists' interests, as the old.[3]

Within those organizations, however, leadership often evolved to become friendlier to the new conception of the field. NSF, for example, saw a director appointed who was much more closely attuned to the economic impact of science, and who created programs that supported university–industry collaboration (US House, 1978). The agency also saw a growing number of engineers move into leadership positions, which further aligned it with the new model (Belanger, 1998). Similarly, while the new conception did not originate with university leadership, some university administrators, like Harvard's Derek Bok, embraced it particularly early, and used it to their advantage (Bok, 1981).

Incumbents had some ambivalence to parts of this shift. For example, the American Association of Universities chose not to support the 1980 Bayh–Dole Act, which allowed universities to patent their inventions, because it was unsure whether universities should be involved in patenting (S. Steinbach, interview by author, March 31, 2005). But in general, the new settlement did not take hold when challengers displaced incumbents, but when incumbents accepted a new role for academic science because doing so was preferable to allowing the unsettled period to continue.

Effects on Other Strategic Action Fields

This new settlement was institutionalized in a variety of ways. One was through laws: the Bayh–Dole Act, for example, allowed universities to patent their inventions and encouraged them to do so, based on the assumption that providing the incentive of patent rights would move scientific inventions into the marketplace more quickly (Berman, 2008). Another was through funding patterns. NSF, as just mentioned, began to support university–industry research in the late 1970s but also expanded its support for engineering dramatically during the 1980s (Belanger, 1998). Individual states also began to support university–industry collaboration with the goal of driving regional economic development (Berman, 2012a, pp. 135–139). Yet another vehicle for institutionalization was universities' acceptance of new kinds of activities, like faculty entrepreneurship and industry collaboration. While such practices were once actively discouraged in academia, universities came to value their role in connecting academic research with commercial applications, in no small part because policymakers were calling for more such connections (Berman, 2012a, pp. 154–156).

But the new settlement in the field of US academic science also had external effects. It created new fields: the expansion of technology transfer activities within universities, for example, gave birth to a field of professionals devoted to managing such activities and who had an interest in promoting them. The creation of state programs that supported university–industry collaboration with the hope of achieving economic development spawned strategic action fields organized around the question of whom such programs would fund, and to what ends. One might even credit the new settlement with launching the entire biotech industry.

The new settlement also, as F&M would predict, disrupted adjacent fields to a certain extent. For example, because academic science was moving toward closer relations with industry, the young biotech field tended to blur the boundaries between the two. As a result, biotech retained many of the values of academic culture, such as relative openness and a flatter organizational hierarchy (Smith-Doerr, 2004). This shift in values in turn seeped into overlapping fields like the pharmaceutical industry, contributing to what Kleinman and Vallas (2001) have called the "asymmetric convergence" of the cultures of university and industry science.

The cycle of disruption and settlement that took place in US academic science between the late 1960s and the early 1980s did not result from a rupture as sharp or as widespread as the one seen in the field of US racial politics or the US mortgage market, the cases that F&M use to illustrate their approach. Nor did the cast of players change so radically. But the basic dynamics they identify – of external shocks disrupting the old settlement, of a period of crisis and contention, and of a new settlement emerging and bringing relative stability to the field – can clearly be seen at work.

Nevertheless, there are aspects of this transformation that F&M's framework overlooks, or cannot account for. Specifically, the new settlement in

US academic science cannot be explained as the result of challengers taking advantage of a period of crisis to remake the field's rules in ways that favored themselves. It also suggests a better-defined transition from crisis to settlement, and a clearer resolution of contention, than actually took place in this case. In the section that follows, I return to my own field-based analysis of this case, using it to highlight these other parts of the story, and to sketch another way of thinking about the causes of change.

EVOLVING INSTITUTIONAL LOGICS AND PRACTICES IN THE FIELD OF US ACADEMIC SCIENCE

F&M believe that field-level change typically happens when a group of challengers takes advantage of a period of turmoil, itself usually caused by events external to the field, to promote a new conception of the field and new rules governing it that will favor themselves over the previous incumbents. But while some groups tried to take advantage of turmoil in US academic science to forward their own interests in specific ways, they did not try to reshape the field as a whole. Nevertheless, the field *was* ultimately reshaped. How, then, do we explain this change, if it was not the result of strategic action by challengers?

I approach this question by focusing on the coevolution of institutional logics, local practices, and the field's larger environment. From this perspective, the central change in US academic science is the growing influence of the institutional logic of the market, which sees science in terms of its economic value, relative to the institutional logic of science, which sees knowledge as important for its own sake. I assume that these logics coexist but are instantiated in specific practices which may be more or less common or successful in a particular field at a particular moment.

Focusing on the prevalence of competing logics as instantiated in practices provides another way to think about field-level change in cases where there is no obvious project to promote a new conception of the field or new rules governing it. The previous section indicated that a new conception of the field of US academic science came from the policy field. But that narrative is still blurry around the question of how policymakers' beliefs about the economic importance of science and technology actually impact the field of academic science. In the rest of this section, I renarrate this story using the language of logics and practices, which allows us to bring this process into clearer focus.

The Era of Science Logic and its Close

In the field of US academic science, multiple logics have long been visible. These include market logic, which sees the purpose of science in its ability to create things that have economic value in the marketplace; the logic of science, which values the production of knowledge for its own sake; and the logic of the state, which sees science's worth in the service it can potentially provide to the government.

It is possible to identify some practices based on any of these logics at any moment during the last century. However, science logic was relatively dominant during the period of stability and expansion sketched in the last section as the "postwar settlement." The logic of the state was also quite visible during this period, but practices based on the logic of the market could be seen as well (Berman, 2012a, pp. 26–29). Nevertheless, the prominence of science logic meant that scientists had a great deal of autonomy in setting their own research agenda and priorities during this period, and could count on government support to expand from year to year.

Faculty who participated in science-logic practices, like seeking out grants to support curiosity-driven research, found the environment very favorable. Funding was rapidly increasing at a variety of federal agencies, and even the more applied agencies were quite friendly to basic research. The atmosphere was supportive for universities that wanted to expand their PhD programs, for example, or to reduce teaching loads so that faculty could devote more time to self-directed research. Both cultural attitudes and the resource environment made science-logic practices easy to reproduce and to expand.

At the same time, market-logic practices, including research parks, industrial affiliates programs, and industrial extension offices, were all visible and spread to a certain extent. But while they did not, in fact, experience the cultural resistance one might expect in an environment dominated by the logic of science, they did suffer from a scarcity of resources that could potentially support them, and such practices did not thrive (Berman, 2012a, pp. 23–35).

As already discussed, however, by the late 1960s this period was coming to a close as the deal between universities and the state that it rested on became increasingly untenable. By the early 1970s, it was harder to find resources for science-logic practices. Observers were declaring a "recession" in academic science, and the path out of that recession was unclear (Boffey, 1970; Semas, 1973).

The Policy Field Changes the Resource and Regulatory Environment

In the late 1970s, the resource and regulatory environment of academic science would change in ways that would start to favor market-logic practices instead. This happened, as suggested in the previous section, when policymakers, encouraged by representatives of large R&D-intensive companies, came to embrace the idea that technological innovation was critically important to the economy. But while this argument gained bipartisan popularity during a time of economic malaise, what policymakers should *do* to encourage innovation was less clear. Many policy possibilities were loosely compatible with the idea of improving innovation, and people of varying political stripes proposed changes ranging from the highly interventionist to the strongly free-market. By late 1979, "seventy-five or eighty" bills purporting to improve technological innovation were under consideration by Congress (Innovation, 1979).

The popularity of the innovation idea did not result in a coherent new policy direction. But it did give a boost to a variety of proposals from across the political spectrum that could somehow be argued to strengthen technological innovation. These ranged from decisions about what court should adjudicate patent lawsuits, to what the tax rate should be on capital gains, to whether pension funds could invest in venture capital (Berman, 2012a, pp. 159–161; 2012b, pp. 278–285). Some of these, like the Bayh–Dole Act, which allowed universities to patent government-funded research, directly targeted universities. But most did not. Nevertheless, by encouraging activities that treated scientific research in terms of its economic value, their collective effect was to change the broader environment in ways that encouraged the growth of market-logic practices in academic science.

Market-Logic Practices Spread in US Academic Science

In this new environment, such practices began to thrive. Activities like faculty entrepreneurship in the biosciences and the patenting of university research became common, and organizational forms like the university–industry research center diffused. While each of these practices predated policymakers' interest in encouraging technological innovation, they had, in the past, encountered significant barriers that limited the extent of their spread. Would-be biotech entrepreneurs lacked venture capital, patenting of government-funded research was restricted, and little support was available for university–industry collaborations (Berman, 2012b, pp. 273–277).

But the changed policy environment removed those barriers. For example, while breakthroughs in recombinant DNA technology took place in 1973, and the first modern biotech company, Genentech, was founded in 1976 by a UCSF professor and a venture capitalist, scientists had started only a few small biotech firms by 1978. Despite the obvious promise of the technology, such entrepreneurship was restricted: less by the cultural objections to faculty entrepreneurship, though those did exist, than by impending legislation that would strictly regulate recombinant DNA research for safety reasons, and by an abysmal venture capital environment (Berman, 2012a, pp. 62–69).

Those limitations were removed, however, by a variety of policy decisions, each of which was shaped by the new attitude toward innovation that was sweeping Washington in the late 1970s. Proposals to restrict rDNA research, which were seen as "inevitable" in early 1977 (Academy Opposes, 1977), were removed from the table as the debate was reframed around the question of whether such regulation would stifle an emerging high-tech industry (Berman, 2012a, pp. 62–69). The venture capital environment was transformed by new rules that allowed pension funds, which controlled $300 billion of capital, to invest some of that into new ventures (Board of Governors, 2007, p. 67), and by a large cut in the capital gains tax (the top rate was lowered from 49% to 25%). These decisions, too, were shaped by new arguments about "the necessity of a broad, coordinated approach to improving innovation" – in this case, by providing an environment that was friendlier to risk capital (US Senate, 1979, pp. 1–2).

The result of this change in the resource and regulatory environment was an explosion in the number of faculty founding biotech firms and in the massive growth of the small handful that had already been established. By mid-1980, the four oldest startups had a paper value of an astonishing $500 million (Wade, 1980), and after Genentech's late-1980 IPO, its cofounder, UCSF biochemist Herbert Boyer, was worth $65 million (Cole, 1980). By the end of 1981, at least 80 biotech startups had been created (US Congress, 1984, p. 93), with each firm typically involving academic scientists not only as cofounders but as members of scientific advisory boards and consultants (Kenney, 1986).

The case of biotech entrepreneurship is illustrative of how policymakers' new ideas about the economic impact of technology innovation changed the environment of academic science in ways that encouraged market-logic practices. But it is not the only example. A similar pattern can be seen in the spread of university patenting and in the takeoff of university–industry research centers (Berman, 2012a, pp. 94–145). And even market-logic practices that had existed since the 1950s and 1960s but limped along for lack of adequate resources, like research parks and industrial affiliates programs, found themselves rejuvenated by the new appreciation for science that could demonstrate an economic impact (Berman, 2012a, pp. 150–151).

As restrictions on market-logic practices – both legal and cultural – declined, and as more resources became available for science done with the purpose of realizing economic value, market logic became both more visible and more widely used in the field of academic science. While science-logic practices were still quite common, as were those based on the logic of the state, their relative influence declined as it became more challenging to find the resources to support them. The multiple logics would continue to coexist and compete as they had in the past, so calling the shift one from settlement, through crisis, to new settlement seems to overstate the extent of the change. Nevertheless, by the end of this period a new balance among these multiple logics, one that would prove relatively stable, was established and persists to the present.

TOWARD A (MORE) UNIFIED FIELD THEORY?

Testing the ability of F&M's framework to explain a familiar empirical case demonstrates both its value and its limits. Doing so in conversation with an explanation based on the coevolution of institutional logics, practices, and environments can help address the blind spots of each approach.

F&M's theory brings additional value to organizational explanations of field change. Their attention to the relationship between fields, which goes beyond most of organizational field theory, is particularly welcome. If one accepts F&M's argument that the relationship between fields is the most common source of change, as a crisis in one field ripples out into all the others it is linked to (F&M, 2012, p. 19), directly addressing that relationship is critically important. Their framework also returns much-needed attention to power and resources. While power has always been central to Bourdieusian field theory, it too often disappears from organizational field theory, with its cognitive-cultural focus.

The case of US academic science, though, also points to some limits to F&M's explanatory framework. In particular, they focus on the relationship between challengers and incumbents as the proximate cause of field-level change. While they do not say that field-level change *must* result from challengers trying to impose a new settlement on a disrupted field, both cases that *A Theory of Fields* examines in detail reflect such a process. In the field of US racial politics, the civil rights movement was promoting a new vision of legal racial equality that eliminated segregation. In the field of the US mortgage market, the government-sponsored enterprises (GSEs) Fannie Mae and Freddie Mac were trying to encourage a reconceptualization of the mortgage market that was based on securitization (F&M, 2012, pp. 114–163). But in academic science, change did not result from such a strategic, intentional, field-level project, and F&M do not tell us how to explain change in such cases.

F&M's focus on cases of fairly dramatic field transformation – on the transition from settlement, through crisis, back to settlement – also tends to downplay the extent to which the ongoing contestation within fields can lead to evolutionary but meaningful change. Here, there is a tension within *A Theory of Fields*. On the one hand, the authors clearly want to emphasize the constant jockeying and politicking that go on within any field. On the other, they write about very dramatic cases of field-level change and do not tackle the dynamics of this ongoing, more-or-less routine struggle for advantage head on. The changes in academic science were not nearly so dramatic as those in the field of US racial politics or the mortgage market, nor were the boundaries between "old settlement," "crisis," and "new settlement" very clear. Yet shift within the field was nevertheless important, and merits explanation.

Focusing on the coevolution of logics, practices, and environment, though not without limits of its own, is one way to sidestep these issues. It is less a competing explanation for field-level change than a way to explain a different type of field-level change. In fact, I actually began my study of the changes in US academic science looking for a challenger analogous to the ones F&M describe, and it was my failure to find one that led me to other explanatory strategies (see Berman, 2012a, pp. 8–12, for a discussion). My inability to locate field-level challengers caused me to look for other factors that encouraged the development of practices associated with the new regime. What I found was that a diverse group of policy decisions was important in encouraging practices like patenting and faculty entrepreneurship, but they were all shaped by policymakers' new level of concern with the economic impact of technological innovation.

Working backward, I discovered that this concern had increased due to the efforts of representatives of large R&D-intensive companies. Those efforts changed the policy environment in ways that encouraged policies that tried to use science and technology to drive the economy. This, in turn, shifted the resource and regulatory environment of academic science in ways that allowed the bottom-up growth of small, local innovations that leveraged science for explicitly economic purposes.

Thus change that was driven by factors external to the field of academic science – resulting from a project initiated by, and aimed at, members of a different field – affected the environment in ways that facilitated the bottom-up evolution of academic science. But because the change was not intended, F&M have no way to address it and cannot help in locating its source.

This actually helps to answer a question that F&M clearly think is important: how ripples from a strategic action project in one field can contribute to the transformation of another. Field-level change is not only the result of challengers' efforts to change fields. It can also result from the collective effects of local innovations in the context of a changing environment. F&M's incumbent/challenger explanation of field-level change effectively describes one type of transformation, as their empirical cases illustrate. But US academic science suggests another type: cumulative, bottom-up change in practices resulting from the intersection of local innovations and shifts in the field's larger environment. The question of which path is more common, or under what circumstances each one can be found, is largely empirical.

I suspect that the dynamic visible in US academic science may be relatively common. For example, F&M address the transformation of the US mortgage market, but consider a slightly speculative account of the transformation of the field of US student loans. Here, too, what was once a small, stable market for government-backed loans has become a gigantic industry with public and private branches organized, like the mortgage market, around securitization. To the best of my knowledge, nobody set out to transform the field of student loans in the same way that Fannie Mae set out to transform the mortgage market. Yet the field of student loans also suffered from the same external disruptions, in the form of stagflation and a weak economy, as the mortgage market did in the late 1970s and early 1980s.

If the GSEs and investment banks were inventing securitization and related practices like tranching in the nearby field of the US mortgage market, and if they were trying to change policy to remove limits on securitization, then the changes could also have affected conditions for actors in the field of student loans. Actors in the student loan field, probably also experimenting with securitization, would have found that legal barriers had been removed. Their experiments, then, would have been more successful and could have led to a new settlement in the field of US student loans, even if no one actively sought to restructure that field. This account is hypothetical but suggests how the dynamics visible in academic science might also be visible elsewhere.

Thus greater attention to the coevolution of logics, practices, and environments can help us to understand a broader spectrum of field-level changes. This suggests in particular three topics worth exploring more fully as part of an effort to flesh out the connections between a logics-and-practices approach and F&M's theory of fields.

First, we should better specify *how* change in one field is tied to change in a subfield or adjacent field. While I have criticized F&M's overemphasis on strategic action as an explanation for change, I perhaps underemphasize it in a particular way. Though I explain field-level change as the cumulative result of the expansion of local practices, I do not pay much attention to how the originators of those practices were themselves acting strategically in subfields, or attempting to create new fields. While these innovators were not trying to transform academic science as a whole, they did have strategic goals: creating a subfield of university technology transfer, establishing a field of biotechnology firms. It would be useful to think more explicitly about how strategic action aimed at changing a subfield, or establishing a new field, can affect a superordinate field.

Second, we should identify the conditions under which one sees a challenger-incumbent path to change versus change resulting from the spread of new practices. The change in academic science, for example, did not involve a complete restructuring of the field or the replacement of its major players. Might bottom-up evolution, then, be associated with more moderate episodes of change, while more dramatic transformations result from intentional projects to reshape a field? Could bottom-up change typically be an indirect effect of an intentional project in an adjacent or superordinate field? More broadly, are there completely different paths to field transformation not captured by either of these two dynamics?

Finally, we should think about how challenger-incumbent and bottom-up change might coexist and interact. In academic science, challenger-incumbent dynamics did not to explain why the field changed as a whole. But sometimes bottom-up evolution can coexist with projects aimed at changing the field. For example, in the case of US racial politics, F&M note how the mid-1950s rupture in the national field both encouraged the Montgomery bus boycott and also gave it great importance. Doubtless the civil rights leaders who organized the boycott wanted change in the national field. But they were acting much more locally, though in ways that had broad impact because of the field's larger environment (F&M, 2012, p. 127).

In conclusion, field theory is an approach with great potential for the political sociology of science, and F&M's version of it is a real contribution, particularly by virtue of its focus on the relationship between fields and its insistence on attention to power as well as culture. But as US academic science shows, its explanation of how fields change cannot be applied to every empirical case. While F&M have some valid critiques of institutionalism-based organizational field theory, recent work along these lines addresses the complexity and multiplicity of the logics that organize fields while also offering space to address power and resources. Drawing on the strengths of both approaches has the potential to produce something more generative for the political sociology of science, and for sociology more generally, than either approach on its own.

NOTES

1. I am bracketing F&M's explanation for the emergence of new fields, although that explanation is a significant part of their larger project.
2. Internal governance units are organizations "charged with overseeing compliance with field rules" (F&M, 2012, pp. 13–14) and serve as the liaison between one strategic action field and another, particularly the state (p. 78).
3. The Association of University Technology Managers, or AUTM, represents professionals who deal with universities' intellectual property, spinoff companies, and other technology transfer activities.

ACKNOWLEDGMENTS

Many thanks to Mathieu Albert, Neil Fligstein, Scott Frickel, David J. Hess, Dan Hirschman, and Abby Kinchy for their useful feedback and suggestions, and to participants in the June 2012 Political Sociology of Science Workshop at the University of Wisconsin for providing the impetus to write this paper. This work was supported in part by the Richard B. Fisher Membership of the Institute for Advanced Study.

REFERENCES

Academy opposes local DNA research rules. (1977). *Science & Government Report*, May 15, p. 8.

Albert, M., & Kleinman, D. L. (2011). Bringing Pierre Bourdieu to science and technology studies. *Minerva*, *49*(3), 263–273.

Albert, M., & McGuire, W. (2014). Understanding change in academic knowledge production in a neoliberal era. In S. Frickel & D. J. Hess (Eds.), *Fields of knowledge: Science, politics and publics in the neoliberal age* (Vol. 27, pp. 33–57). Political Power and Social Theory. Bingley, UK: Emerald Group Publishing Limited.

Belanger, D. O. (1998). *Enabling American innovation: Engineering and the National Science Foundation*. West Lafayette, IN: Purdue University Press.

Berman, E. P. (2008). Why did universities start patenting? Institution-building and the road to the Bayh–Dole Act. *Social Studies of Science*, *38*(6), 835–871.

Berman, E. P. (2012a). *Creating the market university: How academic science became an economic engine*. Princeton, NJ: Princeton University Press.

Berman, E. P. (2012b). Explaining the move toward the market in academic science: How institutional logics can change without institutional entrepreneurs. *Theory and Society*, *41*(3), 261–299.

Board of Governors of the Federal Reserve System. (2007). *Flow of funds accounts of the United States: Annual flows and outstandings*, 1975–1984. Washington, DC: Federal Reserve.

Boffey, P. M. (1968). Budget paradox: Spending holds even, yet researchers are hurt. *Science*, *162*(3851), 340–342.

Boffey, P. M. (1969). Campus unrest: Riots bring danger of punitive backlash. *Science*, *164*(3876), 161–165.

Boffey, P. M. (1970). Recession in science: Ex-advisors warn of long-term effects. *Science*, *168*(3931), 555–557.

Bok, D. (1981). Business and the academy. *Harvard Magazine*, May–June, pp. 23–35.

Bourdieu, P. (1981 [1975]). The specificity of the scientific field and the social conditions of the progress of reason. In C. Lemert (Ed.), *French sociology: Rupture and renewal since 1968* (pp. 257–292). New York, NY: Columbia University Press.

Bourdieu, P. (1988). *Homo academicus*. Stanford, CA: Stanford University Press.

Bourdieu, P. (1995). *The rules of art: Genesis and structure of the literary field*. Stanford, CA: Stanford University Press.

Bourdieu, P. (2000). *Pascalian meditations*. Stanford, CA: Stanford University Press.

Bourdieu, P. (2004). *Science of science and reflexivity*. Chicago, IL: University of Chicago Press.

Bourdieu, P., & Wacquant, L. J. D. (1992). *An invitation to reflexive sociology*. Chicago, IL: University of Chicago Press.

Boyer, H. W. (1994). Recombinant DNA research at UCSF and commercial application at Genentech. In S. S. Hughes (Ed.), *Oral history conducted by S. S. Hughes*. Berkeley, CA: Regional Oral History Office, The Bancroft Library, University of California.

Camic, C. (2011). Bourdieu's cleft sociology of science. *Minerva*, *49*(3), 275–293.
Cole, R. J. (1980). Genentech, new issue, up sharply. *New York Times*, October 15, p. D1.
Cooper, M. H. (2009). Commercialization of the university and problem choice by academic biological scientists. *Science, Technology & Human Values*, *34*(5), 629–653.
DiMaggio, P. J., & Powell, W. W. (1983). The iron cage revisited: Institutional isomorphism and collective rationality in organizational fields. *American Sociological Review*, *48*(2), 147–160.
Dobbin, F. (2008). The poverty of organizational theory: Comment on: 'Bourdieu and organizational analysis'. *Theory and Society*, *37*(1), 53–63.
Emirbayer, M., & Johnson, V. (2008). Bourdieu and organizational analysis. *Theory and Society*, *37*(1), 1–44.
Endicott, K. M., & Allen, E. M. (1953). The growth of medical research 1941–1953 and the role of public health service research grants. *Science*, *118*(3065), 337–343.
Fligstein, N., & McAdam, D. (2011). Toward a general theory of strategic action fields. *Sociological Theory*, *29*(1), 1–26.
Fligstein, N., & McAdam, D. (2012). *A theory of fields*. Oxford, UK: Oxford University Press.
Frickel, S., & Moore, K. (Eds.). (2006). *The new political sociology of science: Institutions, networks, power*. Madison, WI: University of Wisconsin Press.
Friedland, R., & Alford, R. R. (1991). Bringing society back in: Symbols, practices, and institutional contradictions. In W. W. Powell & J. P. DiMaggio (Eds.), *The new institutionalism in organizational analysis* (pp. 232–263). Chicago, IL: University of Chicago Press.
Geiger, R. L. (1993). *Research and relevant knowledge: American research universities since World War II*. New York, NY: Oxford University Press.
Gibbons, M., Limoges, C., Nowotny, H., Schwartzman, S., Scott, P., & Trow, M. (1994). *The new production of knowledge: The dynamics of science and research in contemporary societies*. London: Sage.
Greenberg, D. S. (1967). LBJ's budget: Lean fare set forth for research and development. *Science*, *155*(3761), 434–435.
Greenwood, R., Díaz, A. M., Li, S. X., & Lorente, J. C. (2010). The multiplicity of institutional logics and the heterogeneity of organizational responses. *Organization Science*, *21*(2), 521–539.
Greenwood, R., Raynard, M., Kodeih, F., Micelotta, E. R., & Lounsbury, M. (2011). Institutional complexity and organizational responses. *Academy of Management Annals*, *5*(1), 317–371.
Greenwood, R., Suddaby, R., & Hinings, C. R. (2002). Theorizing change: The role of professional associations in the transformation of institutionalized fields. *Academy of Management Journal*, *45*(1), 58–80.
Guston, D. H. (2000). *Between politics and science: Assuring the integrity and productivity of research*. Cambridge, MA: MIT Press.
Hardy, C., & Maguire, S. (2008). Institutional entrepreneurship. In R. Greenwood, C. Oliver, R. Suddaby, & K. Sahlin-Andersson (Eds.), *The Sage handbook of organizational institutionalism* (pp. 198–217). Thousand Oaks, CA: Sage.
Hess, D. J. (2012). *Good green jobs in a global economy: Making and keeping new industries in the United States*. Cambridge, MA: MIT Press.
House, Senate Split on authorization for NSF. (1977). Science & Government Report, May 15, p. 5.
Innovation: Some positive steps. (1979). *Business Week*, November 26, p. 96.

Kenney, M. (1986). *Biotechnology: The university-industrial complex*. New Haven, CT: Yale University Press.

Kevles, D. J. (1978). *The physicists: The history of scientific community in modern America*. Cambridge, MA: Harvard University Press.

Kleinman, D. L. (1995). *Politics on the endless frontier: Postwar research policy in the United States*. Durham, NC: Duke University Press.

Kleinman, D. L., & Vallas, S. P. (2001). Science, capitalism, and the rise of the "knowledge worker": The changing structure of knowledge production in the United States. *Theory and Society*, *30*(4), 451–492.

Lave, R. (2012). *Fields and streams: Stream restoration, neoliberalism, and the future of environmental science*. Athens, GA: University of Georgia Press.

Lave, R., Mirowski, P., & Randalls, S. (2010). Introduction: STS and neoliberal science. *Social Studies of Science*, *40*(5), 659–675.

Leblebici, H., Salancik, G. R., Copay, A., & King, T. (1991). Institutional change and the transformation of interorganizational fields: An organizational history of the US radio broadcasting industry. *Administrative Science Quarterly*, *36*(3), 333–363.

Leydesdorff, L., & Etzkowitz, H. (1996). Emergence of a triple helix of university-industry-government relations. *Science and Public Policy*, *23*(5), 279–286.

Lounsbury, M., & Crumley, E. T. (2007). New practice creation: An institutional perspective on innovation. *Organization Studies*, *28*(7), 993–1012.

McGuire, W. L. (2011). *Constructing quality in academic science: How basic scientists respond to Canadian market-oriented science policy – A Bourdieusian approac*h. Ph.D. dissertation, School of Public Health, University of Toronto.

Mohr, J. W. (2005). *Implicit terrains: Meaning, measurement, and spatial metaphors in organizational theory*. Santa Barbara, CA: Department of Sociology, University of California. Retrieved from http://www.soc.ucsb.edu/ct/pages/JWM/Papers/ImplictTerrains.htm. Accessed on February 12, 2014.

Moore, K. (2008). *Disrupting science: Social movements, American scientists, and the politics of the military, 1945–1975*. Princeton, NJ: Princeton University Press.

National Science Board. (2010). *Science and engineering indicators 2010*. Arlington, VA: National Science Foundation.

National Science Board. (2012). *Science and engineering indicators 2012*. Arlington, VA: National Science Foundation.

National Science Foundation. (2013). *FY 2012 performance and financial highlights*. Arlington, VA: National Science Foundation.

National Science Foundation, Division of Science Resources Statistics. (2003). *Federal funds for research and development, detailed historical tables: Fiscal years 1951–2002*. Arlington, VA: National Science Foundation.

Owen-Smith, J. (2011). The institutionalization of expertise in university licensing. *Theory and Society*, *40*(1), 63–94.

Purdy, J. M., & Gray, B. (2009). Conflicting logics, mechanisms of diffusion, and multilevel dynamics in emerging institutional fields. *Academy of Management Journal*, *52*(2), 355–380.

Richard J. Barber Associates, Inc. (1975). *The advanced research projects agency, 1958–1974*. Springfield, VA: Defense Technical Information Center.

Schneiberg, M., & Lounsbury, M. (2008). Social movements and institutional analysis. In R. Greenwood, C. Oliver, R. Suddaby, & K. Sahlin-Andersson (Eds.), *The Sage handbook of organizational institutionalism* (pp. 648–670). Thousand Oaks, CA: Sage.

Semas, P. W. (1973). "Deep recession" is seen hitting basic research. *Chronicle of Higher Education*, January 8, p. 1.

Slaughter, S., & Leslie, L. L. (1997). *Academic capitalism: Politics, policies, and the entrepreneurial university*. Baltimore, MD: Johns Hopkins University Press.

Smets, M., Morris, T., & Greenwood, R. (2012). From practice to field: A multi-level model of practice-driven institutional change. *Academy of Management Journal*, *55*(4), 877–904.

Smith, B. L., & Karlesky, J. J. (1977). *The state of academic science: The universities in the nation's research effort*. New York, NY: Change Magazine Press.

Smith-Doerr, L. (2004). *Women's work: Gender equality vs. hierarchy in the life sciences*. Boulder, CO: Lynne Rienner Publishers.

Swan, J., Bresnen, M., Robertson, M., Newell, S., & Dopson, S. (2010). When policy meets practice: Colliding logics and the challenges of "Mode 2" initiatives in the translation of academic knowledge. *Organization Studies*, *31*(9–10), 1311–1340.

Swazey, J. P., & Reeds, K. (1978). *Today's medicine, tomorrow's science: Essays on paths of discovery in the biomedical sciences*. Washington, DC: US Department of Health, Education, and Welfare.

Thornton, P. H., Ocasio, W., & Lounsbury, M. (2012). *The institutional logics perspective: A new approach to culture, structure, and process*. Oxford, UK: Oxford University Press.

US Congress, Office of Technology Assessment. (1984). *Commercial biotechnology: An international analysis*. New York, NY: Pergamon Press.

US House, Committee on Science and Technology, Subcommittee on Science, Research, and Technology. (1978). *1979 National Science Foundation Authorization*. Washington, DC: US Government Printing Office.

US Senate, Committee on Government Operations, Subcommittee on Government Research. (1967). *Research in the service of man: Hearings*. Washington, DC: US Government Printing Office.

US Senate, Select Committee on Small Business, Subcommittee on Antitrust, Consumers, and Employment. (1979). *Small business and innovation*. Washington, DC: US Government Printing Office.

Vaughan, D. (2008). Bourdieu and organizations: The empirical challenge. *Theory and Society*, *37*(1), 65–81.

Wade, N. (1980). Cloning gold rush turns basic biology into big business. *Science*, *208*(4445), 688–692.

Walsh, J. (1966). NIH: Demand increases for applications of research. *Science*, *153*(3732), 149–152.

Walsh, J. (1969). Campus unrest: Congress ponders federal sanctions on universities. *Science*, *165*(3888), 46–49.

"THE TIP OF THE DAY": FIELD THEORY AND ALTERNATIVE NUTRITION IN THE US

Kelly Moore and Matthew C. Hoffmann

ABSTRACT

Field theory is waxing in the sociology of science, and Pierre Bourdieu's work is especially influential. His characterization of field structure and dynamics has been especially valuable in drawing attention to hierarchical and center-periphery relations in science and technology, and to the stability and reproduction of science and technology practices. What field theory does less well, however, is to capture the existence of multiple (including marginal) logics around a given sociotechnical object. Nor does it capture the dynamics of a specific logic of neoliberal capitalism in the US: the cultural and economic value of entrepreneurship that emphasizes the continual reconfiguration of social relations, which has its roots in a longer US history of progress-through-reinvention, and is abetted by new technologies designed to continually "update" and remix. Much better at capturing these qualities, we argue, is an institutionalist *theory in which dynamism, not stasis, is foregrounded, and there is room for multiple, contradictory, and non-cognitive logics to co-exist. Using the expansion of "alternative nutrition" in the US, we show that its*

Fields of Knowledge: Science, Politics and Publics in the Neoliberal Age
Political Power and Social Theory, Volume 27, 223–258
Copyright © 2014 by Emerald Group Publishing Limited
All rights of reproduction in any form reserved
ISSN: 0198-8719/doi:10.1108/S0198-871920140000027016

formation took place via the conjunction of parallel streams of social action that encompassed diverse logics and encouraged creativity and hybridity. More generally, variability in field stability and qualities, not static fields, deserve analytic attention.

Dr. Andrew Weil is one of many contemporary purveyors of the promise of unbounded health and vitality via the endless application of scientific ideas about nutrition (Weil, 1995, 1997, 2000, 2007). Weil and his peers, such as Dr. Mehmet Oz and Dr. Loren Cordain (author of *The Paleo Diet*), follow in the footsteps of US nonscientists who have used scientific claims to advocate dietary change as a means to physical, and often spiritual, health and well-being or forms of self-improvement. Sylvester Graham (2009 [1839]), the advocate of vegetarianism, alcohol abstinence, and frequent bathing was one of the earliest health advocates; another well-known proponent, Francis Moore Lappé, author of *Diet for Small Planet* (1971) helped place vegetarianism back into mainstream American culture by illuminating its relationship to environmental sustainability. For much of American history, these advocates were frequently denounced as quacks. In any given decade, only a few attracted many devotees through the sale of books and cookbooks, public presentations, and by word of mouth (Whorton, 2002). Yet today, there are thousands of these purveyors of alternative diets, catering both to very specific and quite broad audiences: there are diets for the vegan athlete, for new mothers who want to "get their body back," and the Daniel Diet for Christians who want to lose weight. No longer spread just through relatively static and slow forms like books, most of these diets are continually "updated" via email messages, smartphone applications, new web content, recipes, and "tips of the day." Moreover, their interlocutors draw on diverse sources of authority, the logics of economics, spirituality, and science. Taken together, their claims and practices are an unstable web of constantly changing and hybrid logics organized around eating for health. Indeed, incorporated into the messaging of these interlocutors is that instability is a normal social state.

There is no center of gravity among alternative nutritionists; although some, like Dr. Weil, become popular and wealthy, every day more purveyors offer new combinations of science and promises of "wellness." The "field of alternative nutrition," if it could be characterized that way, would look more like an ever-changing web than like a well-defined battlefield. The low barriers to entry (one only needs a web site and a book, both cheap to produce) encourage creativity and hybridity. The evolution of alternative nutrition over the past forty years (as we show below, it became much more visible in the 1970s) is not a story of "shocks" and big battles through crafty strategic action by actors with clear preferences and the capacity to see a field in nearly objective terms, as some have emphasized (McAdam & Fligstein, 2012; Venkatesh, 2013). Nor is it a story of the role of rules and laws shaping a field (Albert & Kleinman, 2011), nor of the power of a dominant logic overtaking all others. Its proliferation, we argue, is the result of the co-emergence of neoliberal entrepreneurialism in culture and economy, and political and cultural attention to spiritual and physical health.

The collective character of alternative nutrition, too, is distinct from the tight and circumscribed field that Bourdieu's (1993) analysis of fields implied. In alternative nutrition, producers are producing for consumers, and in the contemporary US, consumers are always in-the-making, that is, one of the ways that profit is generated is through the continual co-configuration of consumer identities and products. Such reshuffling and shifting makes the array of alternative nutrition possibilities look something like a constantly moving kaleidoscope to which new color shades are always being added. Scientific ideas, bodies, spiritual and political logics and advice, and economic logics are mixed together as packages for consumers, and reflected in the ways that producers market themselves, as well.

We come to this discussion as skeptics about the value of field studies as compared to institutionalist approaches. Our goal is to take up some of the unresolved questions about field-based studies of technoscience, and to draw on institutionalist theories and other theories to attempt to fill those gaps. In particular, we emphasize institutionalism's attention to the co-existence of multiple logics, its concern with social change and creativity, the existence of multiple motivations among actors (over and above "winning," which Bourdieu and his followers have emphasized), and field formation rather than field dynamics. These are critical elements in the study of science and technology in contemporary life, especially in a neoliberal era, in which dynamism and change are socially and politically valued.

FIELDS AS SITES OF FIGHTS AND SOCIAL REPRODUCTION

The sociological and STS turn toward studying fields is in part a reaction against the study of social institutions, and the Foucauldian emphasis on diffuse power relations. In the former case, field analysts are jettisoning the early functionalist underpinnings of old- and new-institutionalism that presumed stability, shared norms, and until recently, provided few mechanisms for understanding change via contention (DiMaggio & Powell, 1991; Friedland & Alford, 1991; Lounsbury & Ventresca, 2002). In the second case, analysts drawing on Foucault's work have been attentive to power-as-outcome, but with little attention to the causal role of law and material relations, or to social movements in making meaningful change (Binkley & Capetillo, 2009). Pierre Bourdieu's conceptualization of fields (1993), which starts with analyst-defined objective field positions and the subjective meanings that actors ascribe to possibilities for action within the field, is the most influential in US sociology, and increasingly, in the sociology of science (Albert & Kleinman, 2011), so that is the version we engage here. Analyzing fields through this lens offers possibilities for understanding struggles over power, and the resulting reproduction of domination. Although Bourdieu argued that fields took different forms in different times and places, his interest was neither in overlaps among field logics nor in democratization processes, but rather in understanding what Norbert Elias would call "states" of social life, or the means by which social life is reproduced (Elias, 1969). Because of their value in illuminating inequality, fields are now waxing as conceptual tools, able to capture the topography of poles in an arena of struggle over material and symbolic power through which inequalities are constituted and made durable (Emirbayer, 1997; Go, 2008; Martin, 2003).

Analysts of scientific fields have demonstrated that scientific fields are reproduced through the effects of rules (Kleinman & Kinchy, 2003; Moore, Kleinman, Hess, & Frickel, 2011), ideologies (Hess, 2014), habitus (Albert, Laberge, Hodges, Regehr, & Lingard, 2008), and institutional location (Hess, 2011). Contemporary field analysis depends heavily on *a priori* definitions of fields; indeed, in Bourdieu's original formulation, he advocated that analysts begin with objective arrangements, and thus, avoid using actors' categories. Thus, analysts have identified the global political field (Go, 2008), the scientific field (Bourdieu, 1975), and the fitness field (Maguire, 2008), for example. These ontologies can be especially valuable for seeing the center and the periphery cross-sectionally, and for observing the effects of relatively fast-paced or singular causes of change, such as changes in laws or a change in a political regime. By firmly bounding the subject, the effects of "inputs" can be easily observed.

Yet there is reason to suspect that attention to objective conditions might not capture all field dynamics. Some analysts, for example, have shown that formal rules are sometimes decoupled from practical actions in the scientific field (Bromley & Powell, 2012; Whittington & Smith-Doerr, 2008). Panofsky (2011) helpfully takes apart the presumed homogeneity in scientific fields, showing that forms of capital from one part of the scientific field do not always lead to capital accumulation in other parts of the field. Similarly, Hess (2012) raises questions about the extent to which the structural hierarchy that Bourdieu presumed is the most useful way to study fields, an issue that Savage and Silva (2013) also raise. As Savage and Silva (2013) argue, the not-centers of fields (they are not "boundaries," nor are they "edges") are also sites of invention, even if those inventions do not immediately make it to the center of a field. Attending only to the center, they demonstrate, would force us to avoid scrutinizing heterogeneity and sources of novelty that may not be highly rewarded in the short run (see also Moore & Hala, 2002). It is critical to investigate when and why something else becomes possible, or even thinkable, even though it may not become part of the mainstream, or may be dormant for long periods (Moore, 2008).

Moreover, although Bourdieu called for analysts to attend to the variability of fields, what he did not anticipate, nor emphasize, was that under particular conditions, the very notion of the stable field might be less useful. Bourdieusian field theory is predicated on the idea that habitus sets out what is possible and that domination of the cultural, political and economic terrain is a key aim. Actors know what they can and cannot do, and what sorts of consequences might happen as a result of certain kinds of actions that they undertake to maximize their position relative to others in a field. Yet this assumption leaves out multiple motivations, and the kinds of actions that feminists, institutionalists, critical race theorists, and scholars of social movements have identified: those that can leave actors worse off, materially and professionally, as they pursue goals that might have collective, long-term or other values, but which do not immediately maximize their own position. Scientists who act to promote social justice at a cost to their career, in Bourdieu's logic, would have to be forced into a Procrustean bed in which they were merely attempting to advance their position in a field of scientists. To be sure, analysts of scientific fields attend to contenders who lose (Frickel, Campanella, & Vincent, 2009, for example), but even this sort of analysis is more concerned with "structural" dynamics of winning and losing than the content of claims, or with theorizing action as derived from multiple motivations.

Again, David J. Hess is an exception: in his long-term project on alternative pathways, he makes clear that there are multiple reasons for action within a scientific field, which include something other than dominance over other producers (see, e.g., Hess, 2007). For our purposes, it is critical to reject the assumption that field domination is what motivates actors. It may indeed be what motivates some, but it hardly exhausts the range of possible motivations actors might have, including those that promote greater social goods at a cost to individual actors. Moreover, we want to draw attention to the fact that actors might discover motivations through action, and may or may not be clear about what the range of possibilities for action might be. Social movements can play a role in shifting fields, because at these moments – what Zolberg (1972) called "moments of madness" – relations of power are more visible, and actors begin to articulate new possibilities, with varied and sometimes surprising results.

Finally, although embodiment was a key element of Bourdieu's analysis of systems of domination (1984), most studies of science and technology fields have little to say about embodiment as a target of actors nor about embodiment as a means of knowing. Bodies were critical sites of Bourdieu's analysis of habitus, and are, in the contemporary moment, key sites of technopolitical action. Analysts of neoliberalism have had much more to say about how embodiment becomes a target of political action (Guthman, 2011; Harvey, 1998), and the very vehicle through which actors constitute themselves politically (Bernstein, 2007).

We offer a neoliberal institutionalist view that allows for states ranging between contention and harmony, multiple logics of action, uncertainty, path dependence, and in which centers are only one of the sites of meaningful social action. We turn next to an examination of this framework, and then move to our analysis of the emergence and dynamics of alternative nutrition to illustrate its value.

INSTITUTIONALISM MEETS NEOLIBERALISM

Alternative nutrition providers draw on diverse logics, which we define, following Thornton, Ocasio, and Lounsbury (2012) as "frames of reference that condition actors' sensemaking choices, the vocabulary that they use to motivate action, and their sense of self and identity" (p. 2). Thornton et al. (2012) argue that any given object might be subject to multiple logics, some of which might compete with and some of which might sit awkwardly against each other. Moreover, they decouple logics from institutions, allowing logics to travel through law, through networks (including professional networks), markets, families, and other social relations. In doing so, they break with old and new institutionalisms that tightly coupled logics and institutions. They allow for structures of meaning and action, such as race or gender, to shape action, even if they are not specific to a particular field or institution, and indeed, cross-cut them in various ways. They also challenge Bourdiesian field theory, which treats fields as constituted by defining and narrow logics that limit possibilities for action, and which are given a priori, rather than configured through action.

As a general theory, Thorton et al.'s "institutional logics" perspective is useful for identifying variability in logics as one of many possibilities in studying action around an object or topic (and they leave open the option that some forms of social action might be more homogenous than others). But their framework has a major shortcoming: it is an ahistorical framework. Different geographical and sociopolitical contexts will produce logic mixes – and different cultural values that might shape when and how such mixes could be deployed and how they shape action. The US form of neoliberalism, we argue, encourages the mixing of varied logics, particularly in the marketplace, such that even formerly competing institutional logics have become multiple, intertwined, reinforcing, and complementary. Christianity and sex, for example, used to be counterposed; but today, one can find organizations, manuals, discussion groups, and elaborated languages that advocate that Christians not only have sex, but enjoy it in particular ways.

As a historically and geographically bounded set of cultural ideas and practices, neoliberalism takes many forms and practices (Connell & Dados, 2014; Fourcade-Gourinchas & Babb, 2002; Harvey, 2005). The US form enhances possibilities for hybridization of field logics, not only by infusing economic logics into other fields, but more generally by rewarding innovation, creativity, and opportunity seeking. Thus, it is not simply that everything is turned into a product; it is that entrepreneurship and its purported outcome, innovation, are increasingly rewarded and viewed as the nation's economic "engine." This is not a brand new development in the US, for here, the "new" has always been linked to notions of nationhood and progress. But certainly, the speed with which it is taking place is remarkable, and citizenship itself is increasingly framed in entrepreneurial terms as well, as citizens are called upon to be savvy "consumers" of formerly public and non-economic goods such as health care and education, and pushed to choose the best investments for the greatest returns to retirement, volunteerism (as human capital), education (including "continuing education"), and religious practices (e.g., Jesus as a "life coach").

Not only are citizens asked to take on new risks in this system, they are also called upon continually to seek new information and to use it to manage and revise extant self-investments (Rose, 2007). In terms of human nutrition and health, "investments" are supposed to lead to a long and healthy life. Becoming ill is a failure of individuals, encouraging not only more knowledge seeking but opening up opportunities for the differentiation of products (Metzl & Kirkland, 2010). What is striking about how alternative nutrition purveyors sell their wares is that they do so by linking claims not to scientific facts and dire warnings about what will happen if their advice is ignored: instead, they connect to consumers' bodies via emotion, thus using strategies of engagement far from the distance and sobriety with which scientists normally present their work.

A second historically and geographically specific change, and deeply related to US neoliberal entrepreneurialism, is the rise of Internet technologies. They have, first of all, lowered barriers to entry to information-related practices such as alternative nutrition. Like other technological disruptions, moreover, the Internet promotes innovation (Baron & Shane, 2008) by changing the speed of knowledge circulation, and permitting new knowledges to be circulated in multiple forms (e.g., videos, writing, still images, sound). To return to the kaleidoscope metaphor, it permits the mixing of more kinds of representations, combined with the rapid changeover of any given representation (e.g., one's computer "updates" without being asked).

Contemporary technologies and neoliberal logics of entrepreneurialism mutually reinforce each other, such that the Internet increases the speed of the circulation of ideas as well as goods and allows for formerly distinctive media (film, print, still imagery) to be combined in new ways for new audiences and consumers. These media also offer more possibilities for circulating alternative views from those that dominated former institutional logics, with relatively low costs. Alternative nutrition, which is heavily dependent on the Internet for growth, relies on this kind of mixing, made possible in part by these technologies.

In addition to illuminating the value of an institutionalist approach that takes neoliberal capitalism seriously, our other goal is to make an empirical case about the changing nature of science in public life. After a period of the dissipation of scientists' monopoly on truth about the natural world in the 1970s, science has become more "modernized" with new actors making scientific claims (Hess, 2007; Moore et al., 2011). Making claims in the name of science is no longer nearly monopolized by scientists, yet there is a popularization of scientific knowledge through its linkage to common sense realism (i.e., knowledge is true if it works or makes sense to an individual through practical experience). Alternative nutrition makes individuals' bodies the arbiters of the truthfulness of scientific claims, with significant consequences for the power of science, and scientists. Alternative nutrition purveyors rely heavily on metascientific analysis of extant research, depicting themselves as charismatics who have special abilities to sort through pools of knowledge, and come up with the correct interpretations. They also rely on personal experience, as well, using their own bodies as signifiers of truth, a move that is distinctly different than the traditional efforts of scientists to keep their own bodies outside processes of truth-making, unless they were disciplined through machinery such as microscopes.

Because most alternative nutrition purveyors have web sites that are continually updated, they, too, contribute to the sense that scientific knowledge is unstable and that an entrepreneurial attitude – looking for ever more knowledge to fine tune one's health, a job that is never done – is the key to good health. The embodiment of the practice – literally consuming things that are new or labeled new – deeply instantiates neoliberal ideology and contributes to the destabilization of ideas about nutrition. Thus, although the alternative nutrition purveyors rely on science, they undermine some key elements of its authority, namely, its relatively slow pace, the sobriety and disinterestedness of its practitioners, and its emphasis on law-like (i.e., stable) ideas. The unsettledness of information delivery – everyday updates, new recipes – not only signifies a more porous field but constitutes the instability of scientific knowledge as a durable object and makes it much closer to a commodity whose value is its ability to personally affect us.

We turn, next, to analysis of how and why alternative nutrition emerged as an expansive set of social relations, rather than as an occasional thesis put forth by a charismatic figure. We show that its rise was related to the conjunction of social movements and neoliberal capitalism in the US. We then turn to an examination of the character of two exemplary claimants – Dr. Andrew Weil, of Dr. Weil.com, and Kim Barnounian, one of the authors of the vegan *Skinny Bitch* diet series.

SOCIAL MOVEMENTS AND THE EMERGENCE OF ALTERNATIVE NUTRITION

Before we embark on our analysis of this "field," we want to be reflexive in our treatment of the concept. Identifying a field (or a population, an institution, or a network) can come from extant theory or empirical studies, but it can also be created by investigators to bound an object of analysis. We could call the nutrition interlocutors that interest us part of a field of "self-help," or of a field of "biomedicine," or any number of other fields, but that fact supports one of our key ideas, which is that it is important to attend to the contemporary situation in which some fields, particularly those dominated by professions such as science, are made more porous and malleable. We examine the way that "alternative nutrition" came to be an arena of action characterized by logics drawn from distinct social processes (federal eating policies and the spread of "counterculture" styles of living) and most recently, by a highly unstable set of logics. We illuminate the weaknesses of field theory in explaining this history, and the comparative value of institutionalist approaches.

Forming a Field?

Human nutrition developed as a scientific specialty in the late nineteenth century, beginning with the experiments of pioneering chemist Wilbur O. Atwater. Inspired by European nutrient charts, first as a student at Yale University and then as a professor at Wesleyan University, Atwater undertook the measurement of macronutrients, including fats, carbohydrates, and proteins, and micronutrients. In 1893, Congress allotted funding for studies of human nutrition deficiencies; Atwater took advantage of this funding and developed new studies of the digestibility of food and its use as human body "fuel" (Maynard, 1962).

Despite this early government funding, few of Atwater's analyses were used by anyone other than other scientists, and this pattern continued in the 1920s when the United States Department of Agriculture (USDA) became a permanent home for human nutrition research in the federal government, with food composition as its central activity (Beecher, Stewart, Holden, Harnly, & Wolf, 2009; Levenstein, 2003). The practical application of this new knowledge began in earnest in the 1930s via the USDA Home Economics Division. Led by Hazel K. Steibling, the division translated scientific knowledge into practical guidelines for food relief for the poor (Dupont, 2009). Given the USDA's dual role in promoting agriculture and human nutrition, not surprisingly, many of the food products and food groups about which most information circulated were those overproduced by US agriculture, including wheat, meat, and dairy products (Pollan, 2009). Dieticians, food columnists in women's magazines, schools, and women's clubs were the major vehicles through which this scientific research was disseminated. By the 1950s, human nutrition science was a well-defined (if not prestigious) scientific field, separate from chemistry, its original home. Human nutrition programs were flourishing, particularly at Tufts University, The Johns Hopkins University, and Columbia University. This traditional arrangement of the field, with bench scientists carrying out research funded mostly by the government, and the dissemination of this knowledge via a layer of technical interlocutors, was based on the idea that the problem of nutrition was largely a problem of applying the right scientific standards to individuals.

A Field in Formation Fragments: From Nutrition to Hunger and Back Again

But in the late 1960s, human nutrition came to be a hot political topic, decentering the technical foci and emphasis on human growth and introducing new actors into the field. In 1967, members of the Congress of Racial Equality and other civil rights field organizers organized an investigative trip to draw attention to the problem of hunger among the largely African-American poor southerners. The liberal Field Foundation of Chicago sponsored three members of the US Congress Subcommittee on Poverty, four physicians, and a civil rights attorney to travel through Mississippi and Alabama to document the extent to which American citizens were going hungry. The trip brought them into contact with forms of malnutrition they had only read about. In their report, the doctors wrote:

> We saw children being fed communally, by neighbors who give children scraps of food because their parents have nothing to give. We saw children who were horribly diseased, with obvious signs of malnutrition, including distended bellies and hair loss, and mental lassitude and poor cognitive development. Other children had only partial vision, and many were missing teeth or whose teeth had never come in. It was hard to believe we were examining American children of the twentieth century. (Quoted in Maney, 1989, p. 8)

Based on this report, the Field Foundation and a group of attorneys and poverty activists created a documentary called *Hunger in America*, which aired on CBS TV during prime time to more than 10 million viewers. It used graphic footage and doctors' testimony rather than the cold calculus of numbers to convey its message. The next day, Congress was flooded with calls from outraged citizens who demanded that something be done about the problem. A new subcommittee, the Senate Select Subcommittee on Nutrition and Human Needs (SSCNHN), was organized by Agriculture

Committee members Senator George McGovern (D-SD) and Senator Robert Dole (R-KS) (Kotz, 1971; Maney, 1989).

The Field Foundation trip and the subsequent video placed food back on the nation's political agenda for the first time since the Great Depression, and it included a role for Congress as a mediator of political and scientific claims and concerns. The Senate Subcommittee took testimony from civil rights workers, nutrition scientists, government employees, and many others (although only one poor person was ever asked to speak in the committee's nine-year history), seeking to understand why so many people had too little to eat and why the food that they did have – sometimes via transfer programs like the USDA's surplus food programs – was of such poor quality that it produced hunger and malnutrition. This committee's work was instrumental in generating an extraordinary set of policies to address the problem of too little money and too little food, including increases in food stamps and the establishment of school breakfast programs (Eisinger, 1998; Maney, 1989).

Like other scientists who came to participate in public political debates in the 1960s and 1970s (Moore, 2008), some nutrition scientists were especially eager to join the debate and discussion, and to seek new ways to address the nutrition needs of the poor. Joan Gussow of Columbia University, Neville Scrimshaw of Tufts University, and especially, Jean Mayer of Harvard, and later Tufts, University were particularly important. Mayer chaired the 1969 White House Conference on Food, Nutrition, and Health that President Nixon organized after *Hunger in America*, and was influential in getting members of Congress to support antihunger programs.

Field theorists might treat their participation strictly in terms of competition with others in the field of human nutrition, as an effort to gain status and power over other scientist competitors. But such a view provides a rather thin conception of human motivation and has little place for anything like sympathy. While we do not have any direct evidence of the thoughts and emotions of actors like these, but neither do we hold reductionist assumptions about motivations. Certainly other evidence suggests that these kinds of actions are motivated by more than professional advancement or expectations of massive field change; indeed, for scientists, participation in public political debates may be harmful to them, leaving the Bourdieusian field analyst with few tools, save perhaps that actors made an incorrect reading of the field or lacked skill (McAdam & Fligstein, 2012) to explain this kind of action. The field concept, and the notion of "habitus" may therefore be limited, given their conceptualization of how social life is ordered and how actors participate in it.

When actors undertake projects that defy standard field logics, some hope that others will follow suit, but nonetheless many still participate in these kinds of actions despite considerable professional and personal costs (Moore, 1996). Although to some degree Bourdieusian field theory can accommodate the participation of individuals in multiple fields (and the various types of capital that they acquire and convert across fields), the concepts of field-based capital and habitus may be too restrictive. Institutionalist theory permits actors to have multiple motivations (Clemens, 1997; DiMaggio & Powell, 1991; Sarasvathan, Dew, & Ventresca, 2009; Thornton et al., 2012), and it enables the *discovery* of interests through action. The participation of high-profile nutritionists like Gussow, Mayer, and Scrimshaw opened up new possibilities for action around human nutrition, making the problem of nutrition for the poor a robust subject of study at Tufts and other universities (Maney, 1989). It did not displace bench science relating to the chemistry and biology of human nutrition generally as the engine of the field, but it did help to encourage many other nutrition scientists to participate in a variety of public political debates over nutrition. This "social movementization of professions" (Moore, 2008) was common the 1970s, offering what Hess might call an "alternative" pathway for professionals.

Neoliberalism and the Informationalization and Personalization of American Dietary Advice

This moment of attention to nutrition as a political and economic problem of poverty was short lived. By 1977, a rather extraordinary field shift had taken place: the SSCNHN speakers had little to say about the problem of too little to eat and much more to say about the health consequences of *too much* food and too much of particular kinds of foods for all Americans. The new assumption about the social problem of nutrition was that Americans had enough money but had *deficits of information* about what to eat. Moreover, the new target of eating policy was the citizen at risk of cancer and heart disease, not of malnutrition (Nestle, 1993). In January of 1977, the SSCNHN announced the release of a new set of human nutrition guidelines, *Dietary Goals for the United States,* the first that had been produced in the US since the *Basic 4* guidelines release in 1955 (SSCNHN, 1977). For the first time, federal guidelines called upon Americans to eat *less* of some things, including salt, sugar, and fat, and to *increase* their intake of other foods. This new balancing act was distinctly different from what was encapsulated in earlier guidelines, such as the 1956 *Basic 4* or the 1943 *US Needs Us Strong* guidelines, which both encouraged a foundational diet with minimum servings of each kind of food (Nestle, 2007). In 1979, the American Society of Clinical Nutrition produced another report, sponsored by the Surgeon General's Office, which drew similar conclusions about the American diet, asking Americans to undertake an assortment of monitoring activities to generate health. The new guidelines were different than earlier guidelines for a second reason, too: the problems they were supposed to solve went beyond chronic illnesses and included lassitude in schools, problematic family life, dental problems, low birth weight, and a range of other ideas. The ideas were advanced by policy entrepreneurs who testified before Congress or who had influence in the USDA, whose charge it was to create the guidelines. The federal human nutrition research agenda turned quickly toward studies of nutrition for health. In 1977, a report by the US Government Accountability Office (GAO) – requested by Senator Hubert Humphrey (D-MN) and written by a diverse group of blue-ribbon scientists, health advocates, economists, and dieticians – declared that American nutrition research had done dismal job of identifying the links between eating and diseases, and recommended increased funding (GAO, 1978).

The 1977 guidelines – fully fleshed out for the public in the 1989 *Nutrition and Your Health: Dietary Guidelines for Americans* – are important because they are mark a critical shift in the imagined user of the guidelines. Earlier guidelines presumed that a woman was cooking for a family. But the new guidelines presumed an "average" individual in need of information, not money. Furthermore, the language of the new guidelines included new economic metaphors; for example, "consumers" were to modify what they purchased and what they ate. A new law required that the guidelines be updated every ten years, based on the latest research, a law that some critics have said "confuses" publics, because the basic ideas about what is healthy to eat are neither complex nor novel (Nestle, 2007).

Of course, these debates about nutrition cannot be understood without reference to the broader set of shifts in how disease was conceptualized and how people came to be ill. As Brandt (1997) argues, by the early 1970s an emerging critique of modern biomedicine and medical technology centered attention on the question of responsibility for disease and its prevention. According to influential economic conservatives such as John Knowles of Rockefeller Foundation, there were diminishing returns from technology and tertiary care; instead, the failure was in the prevention of disease. The goal of health and longevity, in Knowles' view, rested firmly with individuals, who in the previous decades had forfeited their health with an "orgy" of greed, avarice, and overeating, the "diseases of affluence." Knowles called for a return to the Puritan values of self-discipline and moral restraint. Eager to reduce the "dole" implicit in rising health expenditures, Knowles suggested that the "idea of a 'right' to health should be replaced by the idea of an individual moral obligation to be preserver of one's own health … a public duty if you will" (quoted in Brandt, 1997, p. 90). Disease, in other words, was no longer a random event but a failure to take precautions against publicly defined risks. By the 1970s, Knowles' ideas had reached the mainstream: in 1978, Joseph Califano, Director of Health and Human Services, said, "We are killing ourselves by our own careless habits" (Brandt, 1997, p. 90). What these habits were would have to be unpacked not only by scientists, but by the critics of mainstream nutrition.

Individuation and Personalization with a Smile: Nutrition as Lifestyle

The flourishing of the counterculture in the US emerged in parallel with these events. Rising prosperity made it possible for explorations of self, aesthetics, and sexuality, normally available mainly to the bourgeoisie, (i.e., "bohemians") to be undertaken by a much wider variety of people. One of the most important features of this period was the flourishing of scientific forms of self-fulfillment, whose apotheosis was what came to be known as the Human Potential Movement (HPM) centered at the Esalen Institute in California. The secular answer to spiritual quests, it was inspired by psychologists Abraham Maslow, Carl Rogers, and Rollo May. The HPM saw endless potential in human beings for creativity and self-expression, after their basic needs were met, and encouraged people to use quasi-scientific principles to improve their lives (Goldman, 2012; Puttick, 2004).

The revolutionary aspect of the HPM and of the broader cultural ethos of "loosening" – ridding oneself of the strictures of convention – was the new form that the relationship between experts and citizens took (Binkley, 2007). At least, "loosening" was advice given to men; for women, "looseness" in mind and body has significant negative consequences, an issue that is not taken up by most analysts of this period. At the end of a particularly technocratic era in American history, universalistic scientific advice, delivered from an Archimedean point of nowhere and everywhere, was discredited. Science remained a powerful cultural and political tool, but scientists qua scientists had been discredited as objective purveyors of scientific information. Scientific authority and scientific claims had become unbound from each other (Moore, 2008). The new form through which scientific claims were delivered was personalistic – the speaker and the recipient were presumed to have a certain kind of intimacy (Binkley, 2007; Kripal, 2007). Thus, expertise in the HPM was delivered not in universalistic terms, but through people with personalities, gurus whose charisma and experience were the filters for the production and assessment of scientific knowledge. As scientists' broad claims to scientific authority were eroding there was an expansion of "experts" and expertise in culture, such that there was a widening number of scientific claims via populist projects such as retreats, special courses, and classes – even as these purveyors of knowledge were often portraying themselves as "anti-experts" (see also Lave, this volume).

In the alternative health field *wellness* came to dominate discourses and practices. Originally developed in the 1950s, its major proponent, James Dunn, described wellness as follows:

> [A]n integrated method of functioning which is oriented toward maximizing the potential of which the individual is capable. It requires that the individual maintain a continuum of balance and purposeful direction within the environment where he is functioning. (Dunn, 1961, pp. 4–5; quoted in Miller, 2005)

As part of this shift, exercise came to be seen as both important for mental and physical "well-being" but also as a pleasurable individual activity. Running, which grew enormously in its popularity during the 1970s, is perhaps the quintessential. The New York Marathon, for example, had 300 runners in 1972; in 1979, more than 3,000 participated in the race (Whorton, 2002, p. 180). The popularity of Kenneth Cooper's concept of "aerobic exercise," introduced in his 1968 book of the same name, helped to fuel this rise (Cooper, 1968).

Perhaps even more significant for understanding the conjoining of science, pleasure, and the body was the development of scientific methods of personally tracking one's exercise. Recording distances, times, and "personal bests" became commonplace in books about sports pursued for fitness. Jim Fixx's *The Complete Book of Running* (1977), for example, advocated such record keeping, and so, too, did other competing running books. Cooper's book told readers that regular exercise was critical and specified the frequencies with which people should engage in it. All manner of books appear on other individual level sports, with instructions on how to engage in them. Colin Fletcher's *The Complete Walker* (1971), for example, told readers how to outfit themselves and stay alive and well on extended backpacking trips, formalizing and adventure into the wild with advice on how to experience nature safely, rather than to be surprised by it. Gyms, which used to be spaces for working-class men who boxed or were part of the esoteric culture of bodybuilding, attracted a new, middle-class (and at first largely male) membership, who participated in bodybuilding using not just free weights but "weight" machines, as individualized practice that was part of an overall "fitness" regime. Gyms served as training places for the middle classes to learn how to move their bodies to "burn calories" and other outputs that were now expected to take up Americans' leisure time (Maguire, 2008).

Counterculture Nutrition as Politics: The Challenge to Industrial Food

Challenges to the industrial food system from the counterculture took a similar form. Some Americans came to believe that urban and modern life, including food, were too corrupt and alienating, and they went "back to the land" to raise their own food. Critiques of food as "plastic," chemical-laden, and non-nutritious became more commonplace. Environmentalists decried the effects of animal grazing on land, and ethical concerns for animals created a surge in vegetarianism. Some of these concerns were organized into food co-ops, but these tended to be short lived (Belasco, 2006). A particular form of feminist nutrition emerged during this time, with the publication of *Laurel's Kitchen* in 1976 by three Bay Area mothers and wives. A lifestyle guide as well as a cookbook – it was subtitled "a handbook" – the authors advocated vegetarianism and home cooking in a communal context. Food was as much about style of living as it was about the proper diet to acquire health, and women were to be the leaders of a new food movement, starting in their kitchens (Robertson, Flinders, & Godfrey, 1976).

Between 1970 and the early 1980s, then, the science of human nutrition in the US, long a cozy world of research scientists focused on minimums and recommendations for citizens who were not ill, began to shift gears. New research emphases were on nutrition for health, and new audiences for that knowledge were set in place by the new nutrition guidelines (which were taught in schools), and by a counterculture that viewed food production and eating as matters of morality and politics. One of the most prominent civil society food organizations, The Center for Science in the Public Interest, founded in 1971, became exclusively focused on problems of nutrition and highly visible in numerous food debates in the 1980s and 1990s. The problems of the poor had largely faded from view by the time that Ronald Reagan was elected president in 1980, but there was a new world of attention to preventing illness and its connection to styles of living.

This set of shifts is not the result of an easily identified event, but the evolution of a field as a result of economic interests of the state and other actors, the development of a sense of civic participation among nutrition scientists and nutrition NGOs like the Center for Science in the Public Interest, and new audiences for particular kinds of lifestyle-related eating and consuming knowledge that fell outside the scientistic frameworks offered by the government. They were unquestionably shaped by a shift in ideologies away from collective problems and solutions to individual harms and solutions, a framework methodically pursued by Ronald Reagan during his Presidency, and by the socialization of some nutrition scientists, who came to look for new ways to use nutrition knowledge to address social problems. Civil society groups, and publics not organized, sought alternatives to supermarket-based industrial food unthinkingly consumed. Not surprisingly, food companies began to label their food as "healthy" and "nutritious," carrying out their own research – but more often, drawing on other scientific work (Schleifer, 2013) – and creating a cacophony of views about the right way to eat. The "field" came to be even more complex and diffuse in the 1980s, as a result of innovations in advertising and publishing – the development of numerical "lists" on covers and very short pieces of copy.

BETWIXT AND BETWEEN: NEOLIBERALISM, SCIENCE, AND THE PLEASURE OF BODY DISCIPLINE

In the 1980s, the number of health and fitness magazines published in the US surged; in 1975, there were five of them, aside from bodybuilding magazines, dominated by the Rodale Press' *Prevention*. *Prevention* was formerly a staid journal giving advice less to counterculturalists than to people interested in their health for religious reasons, for treating illnesses, or because it was good for the land. With low circulation before 1970, the magazine got a boost from the counterculture interest in alternative food. Yet by 1980, in *Prevention* (then and now the best-selling health magazine in the US and the world) stories about neoliberal investment and ownership of one's body and health became prominent. Articles such as "Subtracting additives multiplies your health account" (January 1981, p. 174) and "Now's there's an owner's manual for the most valuable thing you own [your body]" (June 1980, p. 101) were typical. This investment practice came with a call for combinatorial and numericized nutrition: taking ideas, commodities, and foods, and uniquely interpreting them. In "The surge to self-care," (December 1982), editor Robert Rodale wrote an extensive article on the need to seek out and apply a variety of different ideas in order to achieve optimum health.

> But books and magazines and other written sources are only the tip of the self-care information iceberg. Other lay people are an important source of knowledge and support. It is important to realize that self-care, while it may sound at first like something you only do yourself, is really a very social activity. (p. 22)

Rodale urged readers to frequent self-help groups and to produce and distribute their own ideas about "self-care."

In 1982, *Prevention* and nearly every other health and nutrition magazine in the US began to offer information in a new way: as numbered lists and quizzes. Numericizing ideas, as Espeland and Stevens (1998) have shown, has the effect of allowing ideas to be more easily transposable and transferrable. Redacted from sources, they are portable and fungible. Articles with titles such as "Best and worst diet ideas," (January 1985a), "See if you need Vitamin C" (February 1985b), and "Eight reasons you might be going bald" (March 1986) proliferated in the 1980s, calling on readers continually to take bits and pieces of scientific claims via lists of various kinds and to apply them to their own bodies and lives, authoring their own bodily experiences and outcomes with the scientific commodities delivered to them. Since the 1990s, the Internet has work synergistically with this process, allowing new bits of information to be created and circulated instantly.

In the contemporary period, the major players in the field of human nutrition are the federal government, food manufacturers and those who supply their ingredients (seeds, fertilizers, commodities, etc.), nutrition scientists employed by universities, and interlocutors who carry out what we call meta-analysis of scientific research. The federal government creates standards for human nutrition that directly shape what institutionalized groups, such as children and prisoners eat, and it sets voluntary guidelines for what Americans should eat (The Food Pyramid, for example). One of its most important functions is as a gatekeeper for ingredients, limiting or allowing additives, chemically modified forms of seeds, and limiting the volume of imported foods, all strongly influenced by corporations (Bittman, 2008; Nestle, 2007; Pollan, 2007, 2009). In general the field of nutrition has become more fragmented, in part because the problems to be solved morph quite quickly, as researchers redefine nutritional problems on a regular basis, and collaborate across widely dispersed networks (Penders, 2010).

Arguably, among the most critical features of American nutrition are industrial chemicals and processing. Increasingly convincing evidence is linking illnesses to plastic additives like Bisphenol A and other endocrine disruptors (Guthman, 2011), but these issues and the heavily industry-influenced farm bill are hardly discussed publicly, even by stalwart critics of the US food system like Marion Nestle (2007) and Michael Pollan (2007, 2009). The obfuscation of what foods are nutritious is not simply a function of actions of food manufacturers who promise endless health benefits from highly processed products at the same time they pursue chemicalized food production. The voices of individual counterweights such as Nestle, Pollan, and the *New York Times*-recipe-columnist-turned-critic-of-agriculture Mark Bittman (2008), arguably the three most influential civil society critics of the US food system, tend to be lost among the noise of the many who proffer nutrition advice, because of the proliferation and rapid development of highly visible diets promising health and wellness.

All three critics recognize this problem, but blame the government and manufacturers rather than the very "field" itself that has low barriers to entry, rapid knowledge distribution, an orientation toward food as a health product and nutrition as an investment, and in many cases, the offer of alternative nutrition advice through charisma rather than sobriety and rational discipline. It is difficult to see how it might be possible to occupy a center position in the field of alternative nutrition, simply because there are an enormous number of entrants, the materials they have to work with (i.e., science-as-commodity, forms of affect, the Internet, calls for endless investments in the self) are themselves destabilizing. Marion Nestle (2007) comes closest to recognizing the problem through her relentless critiques of the regular production of new federal guidelines for nutrition. Although there is an embryonic alternative food movement in this country, it is too heavily focused on the lifestyle issues of the middle class rather than the reform of the system itself or the problems of the 49 million Americans who are food insecure. Instead, their problems are frequently reduced to merely problems of "food deserts," rather than to what they actually are: money deserts (Alkon et al., 2013).

If alternative nutrition is an objective field, as the logic of Bourdieusian field analysis would insist, it should be possible to identify the centers of power that set the rules of the game, the edges of the field, the contenders for power, and common strategies of action. It would be difficult to do any of that with regard to alternative nutrition in the US, save for identifying logics. Institutionalism best captures the multiple motivations and interpretations, creativity in using codes and rules, and wide variation in the stability of logics of action around particular social objects. Rather than treating the arenas of action under study as objective, already-structured, and heavily coercive, this approach presumes that in most systems of meaning and power, there are leaks, weaknesses, failures, and variations. Importantly, by avoiding attention only to the dominant logics, it can help to see patterns of variation and instability (Adams, Clemens & Orloff, 2015; Clemens, 1997; DiMaggio, 1991; DiMaggio & Powell, 1991; Moore, 2008; Schneiberg, 2012).

To illuminate the value of this approach for capturing the key qualities of contemporary alternative nutrition provision and its variations, we examine commonalities and differences in two popular alternative diets: Dr. Andrew Weil, M.D., and the veganism presented by the *Skinny Bitch* franchise. We examine the convergence of lifestyle, science, and neoliberalism in them, and we show that "styles of living" including elements that are not normally considered in studies that assume just one field logic, are important in explaining what they offer. We also attend to the evidence from web sites, because that technology helps produce the phenomenology of self, scientific, and field instability.

Andrew Weil, M.D.: Guru, Scientist, Investor

Dr. Andrew Weil is a leading figure in the holistic health movement and a major force behind the institutionalization of integrative medicine, which promotes alternative therapies as complementary to conventional medicine. Weil earned his A.B. and M.D. at Harvard University. As an undergraduate in biology he was influenced by the work of the Harvard Psilocybin Project conducted by Timothy Leary and Richard Alpert (Lattin, 2010), and he wrote his thesis on the psychoactive properties of nutmeg. After graduating from Harvard Medical School, Weil interned at Mt. Zion Hospital in San Francisco before spending a year at the National Institute of Mental Health (NIMH) doing drug research. He quit the NIMH after a year, citing the political climate at the time and the subsequent resistance to his marijuana research. Frustrated with the biomedical establishment, Weil began to explore shamanistic healing and started practicing yoga, vegetarianism, and meditation. He was a fellow at the Institute of Current World Affairs from 1971 to 1975, traveling the world exploring medicinal plants and indigenous systems of disease treatment. He did similar work while he served on the research staff of the Harvard Botanical Museum from 1971 to 1984 (Lattin, 2010).

Weil's early work, such as *The Natural Mind: An Investigation of Drugs and the Higher Consciousness* (1998[1972]), included criticisms of American drug policy and explored the effects of drugs on human consciousness. Not until the 1983 publication of *Health and Healing: Understanding Conventional and Alternative Medicine* did Weil introduce himself to the public as a health and healing guru. By this time he had already established a general medical practice and was a part-time clinical faculty member at the University of Arizona, College of Medicine. In 1994, Weil established the Center for Integrative Medicine at the university, where he is still the director, Professor of Medicine and Public Health, and the Jones-Lovell Endowed Chair in Integrative Rheumatology. Weil has also played major roles in two evidenced-based, peer-reviewed academic journals that explore holistic health practices, including *The Journal of Complementary and Alternative Medicine* and the *Quarterly Journal Integrative Medicine*.

Today, Weil has a vast media empire that includes CDs, DVDs, a complex web site, television specials, and ten national best-selling books. His most recent book is a cookbook, *True Food: Seasonal, Sustainable, Simple, Pure* (2012), with recipes based on his "anti-inflammatory diet" developed for his restaurant chain, True Food Kitchens ("globally inspired cuisine"), with six locations in Arizona, California, and Colorado. In addition, Weil has his own line of supplements and a proprietary "vitamin advising" online tool to help individual consumers choose the best combination of supplemental vitamins and oils. There is a "Weil by Dansk" line of kitchen appliances, Dr. Andrew Weil Integrative Footwear, and Weil endorsements of various products including granola bars, salmon sausage patties, skincare products, and the complete line of products from Lucina Italian Organics. Weil is also the Director of Integrative Health and Healing at Miraval Resort, a luxury retreat and spa in Arizona, which hosts the Andrew Weil, M.D. Integrative Wellness Program. The program "incorporates the principles of integrative medicine, lifestyle diagnostic tools, evidence-based methodologies and in-depth consultation to plot your optimal lifestyle and wellness path" (Miraval Resorts, n.d.).

Weil's signature concept is *optimum health*, which takes into account the interplay of spiritual, physical, and social well-being. Weil uses words like wholeness, balance, and resilience to describe what health is. Helping people to improve nutrition is thus only one goal of his work; he also advocates cultivating psychological states, such as those in his book titles, including "spontaneous happiness," "spontaneous healing," "emotional well-being." Like some versions of the alternative food movement of the 1970s, the interplay of eating and spiritual and psychological states is important, but here the knowledge is for sale in a big way: Weil's company, Weil Lifestyle, LLC, oversees nearly a million dollars of revenue each year and supports the non-profit Weil Foundation (Zoom Company Information, 2014). Each month, his web site delivers more than fourteen-million free email messages and receives more than eight-million page views. His books have sold more than ten million copies, and his partnerships are part of his company's goal to distribute "distinctive natural health products and services selected and designed by Dr. Weil" (Weil, 2013a).

Dr. Weil's major dietary focus is on minimizing inflammation because of its relationship to disease. He has his own anti-inflammatory food pyramid to provide guidance to those wanting to achieve optimum health. According to Weil:

> Following an anti-inflammatory diet can help counteract the chronic inflammation that is a root cause of many serious diseases, including those that become more frequent as people age. It is a way of selecting and preparing foods based on science that can help people achieve and maintain optimum health over their lifetime (Weil, 2013b).

Some of the diseases that Weil links to chronic inflammation include cardiovascular disease, various cancers, and neurodegenerative diseases.

Weil's food pyramid, loosely based on the Mediterranean diet, emphasizes fresh fruits and vegetables, legumes and whole grains, and eliminating processed and refined foods. Weil's food pyramid also differs from most food pyramid conceptualizations by including independent categories for fish and seafood, healthy fats (olive oil, nuts, seeds), whole soy foods, cooked Asian mushrooms ("unlimited amounts"), healthy herbs and spices ("unlimited amounts"), tea, supplements, red wine, and healthy sweets. He promotes eating a wide variety of foods in order to cover one's nutritional bases and ensure that one is not getting too much of anything harmful (Weil, 2013b). Such omnivorousness, however, also helps sales among alternative nutrition followers, just as health magazines' encouragement to "eat this, not that," (Zinczenko & Goulding, 2012) emphasizes ongoing learning and consumption of value-added products.

Weil's claims to authority are based in part on his medical background, although he does not draw directly on his own research. Each day on his web site he answers a different question, offering his opinion using a combination of his own views and by sorting through medical research. Weil, like many other alternative diet purveyors, uses a special kind of charisma to convey special knowledge and capacities to sort the scientific wheat from the chaff. Rarely is the information delivered on the basis of purely scientific reasoning; in many cases, he uses his personal experience to justify his answer. For example, in answer to whether black pepper is poisonous, Weil intersperses scientific evidence and his own eating habits:

> I generally don't let waiters grind their pepper mills over my food at restaurants until I taste it first. For a hot spice, I prefer red pepper, which comes from a different plant (*Capsicum spp.*), doesn't have any natural carcinogenic activity, has a long history of medicinal use, and provides healthful carotenoids. It can help lower cholesterol and stimulate circulation, and can actually help heal the lining of the stomach (Weil, 2006).

The relationship between himself as an entrepreneur, a guru, and a biomedical researcher is so entwined in his site's design, the products for sale, and in the messages that are sent out daily that it is difficult to specifically identify the main source of his authority or which logics are operating at any given moment. He moves easily among entrepreneur, spiritual guru, personal guide, and biomedical scientist in ways that undermine the idea of the scientist as sober and distant from their own body and of science as slow and conservative. He does use the logics of science, but the daily web updates and email messages are akin to advertisements that are seen on practically every site on the Internet, enticing us to regularly rethink our future (Zukin, 2008) by updating some aspect of our style of living. (The message-a-day is now ubiquitous; the USDA's Food and Nutrition Information Center issues an email each day, providing diverse bits of information ranging from nutrition research, changes in law, conferences, and policies (fnic.nal.usda.gov). There is no rest for the neoliberal citizen (who has a computer or a smartphone): updating is not optional, but it can be made more palatable with spiritual direction and the right products.

Skinny, Bitchy, and Vegan

In striking contrast to Dr. Weil, whose approach combines a guru-like cheery optimism with scientific credentials and offers something for everyone, the *Skinny Bitch* franchise offers its nutritional, exercise, and lifestyle advice with a tone that is caustic, often demeaning, and vulgar. For example, founders Rory Freedman and Kim Barnouin believe that caffeine addiction is "pathetic" and taunt readers who may want to include dairy products in the diet. To them, the formula is simple: "Healthy = skinny.

Unhealthy = fat." Not only is the desire for meat and dairy products worthy of shame, so too is being fat, regardless of diet. Using the caustic language of humiliation, they refer to being fat as "bloated fat-pig syndrome" (Freedman & Barnouin, 2005, p. 72). Courage and discipline are necessary to participate in their style of living, with high costs for those who are too weak to follow through (and the primary reason for failure to follow through is weakness). This moral call is emblematic of one broader trend in diet and health advice: the return to discipline and toughness as the basis of success. Exercise "boot camps," humili-tainment programs such as *The Biggest Loser* that make contestants run through military-style training regimes (their web site, www.biggestloser.com, urges viewers to lose weight through a series of "personal," "daily," and "team" challenges), and the advice of celebrities like Gwyneth Paltrow, who exercises for two hours each day and urges her followers not to "be lazy," use similar tropes (Dorment, n.d.). Emotional brutality is a poor motivator unless people have options (Rejali, 2009), but paired with "opportunities" for redemption and salvation, it is more powerful (Moore, 2013). By following the *Skinny Bitch* style of living, women can acquire sassiness, confidence, a sense of rebelliousness, and ultimately, the admiration of others but especially of other women. Like other forms of neoliberal advertising that pair economic risk taking with images of happy people, *Skinny Bitch* offers the prospect of being a winner in the competition for health and social admiration. And indeed, they do not offer this as a system in which everyone wins. They draw on a particular form of neoliberal economic logic: winner-take-all, and social shame for the losers (Moore, 2013).

Like Dr. Weil, Barnouin makes claims about her formal credentials. The book jacket notes that she has a "Masters of Science in Holistic Nutrition" (from the non-accredited and now-closed Clayton College of Natural Health). Freedman complements Barnouin by characterizing herself as "a self-taught know-it-all," emphasizing her status, like Weil, as an outsider to the establishment, and placing value on personalism. Although all their scientific claims are presented with citations, the sources are typically articles that come from a news source such as the *New York Times* or from a vegetarian or vegan advocacy organization such as NotMilk.com. Other nutritional claims are traced back to John Robbins' *Diet for a New America* (1987), Harvey and Marilyn Diamond's *Fit for Life* (1985), articles produced by Andrew Weil and associates and by the Center for Science in the Public Interest. The authors are twice-removed from the scientific claims they put forth; that is, their collective scientific credentials are weak, and they rarely cite scientific research directly. Similarly, the basis of their self-help advice draws not on citations to psychology research directly, but on the works of popularizers such as Dr. Wayne Dyer (Ed.D.) and Anthony Robbins (who lacks formal scientific credentials).

In making scientific claims, Freedman and Barnouin's method is to emphasize the chemical properties of food, chemical additives, and chemical reactions that may occur in the body during consumption. They then link these chemicals to a litany of diseases and disorders. For example, they compile a partial list of chemicals that can be found in meat and dairy products, including benzene hexachloride (BHC), chlordane, dichlorodiphenyltrichloroethane (DDT), dieldrin, dioxin, heptachlor, hexachlorobenzene (HCB), and lindane. In the next sentence they claim: "Perhaps that is why eating 'meat' has been linked to obesity; cancer; liver, kidney, lung, and reproductive disorders; birth defects; miscarriages; and nervous system disorders" (p. 110). Similarly, returning to the example of coffee, Freedman and Barnouin claim that human bodies produce fat cells "in order to keep the acid away from your organs." Therefore, "coffee equals fat cells" (p. 113; here, the authors cite the research of another Clayton College of Natural Health alumni, Robert O. Young). To bolster their claims, they sometimes list the research institutions where these studies were conducted, particularly if they are Ivy League institutions.

In the tradition of Robert Owen (1771–1858) and like contemporaries Anna Lappé (*Diet for a Hot Plan*et) and Marion Nestle, Freedman and Barnouin link changes in one's diet to broader social critique. They dedicate a chapter, "Have No Faith: Governmental Agencies Don't Give a Shit about Your Health," to documenting failures of governmental agencies in regulating the food industry. The chapter uses a familiar trope among food activists: the revolving door between industry and the federal agencies that regulate them, specifically, the USDA, the FDA, and the EPA. They mix together metascientific and social justice analysis in other chapters such as "The Dead, Rotting, Decomposing Flesh Diet," "The Dairy Disaster," and "You Are What You Eat." These chapters provide evidence from animal rights and advocacy groups and on the unsanitary and cruel conditions of meat and dairy production and the effects of these foods of health. They also use scientific research to support their claim that animals are "intelligent, emotional, social creatures" (p. 151). The animal victims are thus not accountable for their own state, unlike the presumably fat and unhealthy humans who eat them and their byproducts. In contrast to Weil's approach, Freedman and Barnouin use the political logic of many social movements, which is to gain adherents by making appeals to victimhood with possibility of rebirth through action (Polletta, 2009).

The title *Skinny Bitch* targets women readers; in contrast, *Skinny Bastard: A Kick in the Ass for Real Men Who Want to Stop Being Fat and Start Being Buff* (2009) is an effort to make their diet appeal to presumed male vanity and values. The scientific and moral claims are nearly identical to the sister book, and the rhetoric similarly includes vulgar insults as motivation. In the introduction, Freedman and Barnouin draw on a gendered logic to argue that men do not want to be skinny, but they want to be "buff," – the cheeky popular referent for a fit physique, rather than the older "muscular" or "lean." But they emphasize that this is for "real men" and include one original chapter, titled "No Girls Allowed." This chapter focuses heart disease, erectile dysfunction, and having "the right amount" of testosterone. If male readers are not convinced to adopt a vegan diet, the authors then appeal to men's sports prowess with claims of a vegan diet as the ultimate diet for "athletes," complete with a list of Olympic and professional athletes who ate vegetarian and vegan diets (p. 18). But appeals to "performance" aside, by focusing on being "buff," they call attention to a central goal of their enterprise – making oneself worthy of others' gaze and gaining admiration for that.

The entrepreneurial logic of this approach is evident in a chapter near the end of both *Skinny* books. These chapters are intended to help the reader get motivated and begin setting goals. It covers the "chemical addictions" many have to processed foods and cheese and how to manipulate hormone levels to speed up your metabolism. The authors then list all of the vitamins and supplements that are important to consider in a diet, such as calcium, folic acid, iron, magnesium, omega-3 fatty acids, potassium, B vitamins, vitamins C, D, and E, and zinc. Followed by the chapter "Let's Eat," readers are provided lists of processed organic, vegan foods, organized according to meal (breakfast, lunch, dinner) and alphabetically by the manufacturer's brand name. Many of these brands are chosen based on the authors' testimonies of the items they liked, and are paralleled by similar endorsements peppered throughout the books, including nods to Frey Vineyards organic wine, Eddie's Spaghetti, and Westbrae's organic beans and whole grain pastas.

DISCUSSION: INSTITUTIONAL NEOLIBERALISM

A Venn diagram of the field logics on which the *Skinny* authors and Dr. Weil draw would include (moral and material) economy, biosciences, and personal experience. That Freedman and Bartounin also draw on logics of gender to make their case, however, fits uneasily into theories of "field" logics. It is worth noting, for example, that a leading field theorists Doug McAdam and Neil Fligstein's (2012) analysis of race is about the field of racial politics, not about how racial factors shape action in different fields. In our analysis, the sort of competitiveness (with other women and men) that Bartounin and Freedman advocate could be reduced simply to capitalist logics, yet that misses the power of its gender specificity. The "skinny bitch" is competing against other women for the admiration of other women; the skinny bastard is in competition with other men.

Institutionalist analysis, particularly when combined with critical race and feminist theory, is an easier fit for how sets of action and value such as gender (or race or disability) can be taken into account in analyses of contestation over valued goods. Clemens (1997), for example, does just this, in analysis of the simultaneous emergence of interest groups politics as a form of political action and of how American women came to exercise formal political power. She argues that the usual treatments of political innovation start with those at the center of power; in her study, that would have been those in control of US political parties in the late nineteenth century. Such a strategy would be familiar to field theorists attempting to understand a similar process.

Clemens, however, started from the "outside," by looking at three groups (farmers, labor, women) who were all trying to break the party system in some way or another. She began by analyzing how their social identities and particular forms of organization gave them particular advantages and disadvantages in democratizing American politics. Many of the organizations that she studied, but particularly the women's organizations, advocated greater access to political decision making for Americans other than women, not all of whom could be counted on to act as women's allies. Reducing Clemens' analysis to either a story of the emergence of a new political form or as a story of women's emancipation, misses the co-constitution of these actions, and of the ways that women's gendered interactions with each other and other political actors shaped outcomes. Moreover, Clemens' analysis explains the puzzling fact that as they were organizing, women cared little about leading their organizations per se; when those organizations collapsed, women formed other organizations and did not, as Max Weber would have predicted, maintain the organizations in order to maintain power for its own sake (i.e., a process of goal displacement).

To return to the two diet examples, gender is certainly a strategic tool for Freedman and Barnouin, for it is used implicitly and sometimes explicitly to gain adherents. But at the same time it is not reducible to a strategic tool used to dominate a "diet field"; it is part of the cultural logics from which they work. Barnouin and Freedman come from the highly gendered modeling industry. Gender is constitutive of the ways of seeing the world – including the body, health, animals, and the like – and shapes the capacity of actors to make claims in particular ways. Again, field analysts are likely to insist that gender, or anything else, can be fit into the framework, but without attending to the specific content of social groupings like gender; for example, competition is but one logic of action. A more robust analysis would take into account how social groupings like gender provide tendencies and logics – not fixed interests – that can be valuable in understanding not only the constitution of particular games and contests but how they play out. Because most versions of institutional theory rely on a social psychology of phenomenology rather than rational choice or interest theory, they can capture how and when gender or other logics play out even without self-conscious and deliberate efforts.

A second advantage of institutionalist theory for studying social contestations, such as which diets shall prevail (in the market and in cultural recognition), is that it emphasizes multi-level analysis of successful and failed efforts at social transformations. It also attends to experimentation and learning, deliberate and inadvertent reconstructions of interests, the flexibility of rules and regulations, geographic contexts, and processes of democratization (Adams, Clemens, & Orloff, 2005; Ansell & Vogel, 2006; Leschziner & Green, 2013; Moore, 2008; Schneiberg, 2012; Thelan, 2010). The diet models that are described here are having their heyday. It is not clear, however, whether any one of them is currently victorious or will be in the future; indeed, one can find an enormous range of criticism of all of the diet plans, and of the individuals who advocate for them. Perhaps what is most critical to understand about them is that at this moment, the form that they take draws on multiple and shifting logics, and that this can easily be captured in institutional analysis as a centrally important issue rather than a sideline of the "real" action around diet that occurs at the USDA or Congress.

These alternative diet programs are not simply businesses in which their founders happened to invest; they are representations of entrepreneurial action on the part of founders, action that is defined by mixing logics from biomedicine, American self-reform traditions, neoliberal entrepreneurialism at the level of the founders themselves and of their offerings. To place them in a particular field is of course analytically possible, and we have framed their work in terms of "alternative nutrition." Yet if their own testimony is to be believed, they have mixed logics that have been traditionally separated, through at least two different pathways: a successful medical career promoting alternative health, successful modeling and modeling business careers.

The cases that we investigate are exemplary of the same kinds of mixing that can be observed in the hundreds of alternative diets that are offered every year, offerings that ask users to do something other than regulate the intake of elements of the typical American diet, as traditional nutrition advice does. The sheer number of these diets that are offered at any given time, the frequent updating of the advice by virtue of web technologies, the marketing of various products associated with the diet, and the emotional emphases on diet modification are complex mixtures. On one hand, they are signs of a field that has already formed in the sense of actors who share a common orientation toward an object of action and comprehend the same set of possibilities over which they struggle, and on the other hand, they are also indicators of a weakening field. If we take neoliberal culture seriously, instability is constitutive of the field of action, such that possibilities for action include ongoing disruptions of logics and knowledge stability. That such logics come to be part of embodiment through consumption (consuming ideas and information, purchasing products, and bodily ingestion) makes them all them more powerful.

Attending to the centers of the scientific field and the battles that are waged there have proven valuable to understanding the exclusion of new logics of action, actors, and subjects of attention in science, especially when contestations are overt and explicit (Albert et al., 2008; Hess, 2014). Yet in other cases, the contestation over knowledge claims and who has the capacity to be granted the ability to speak authoritatively have less to do with contention among actors for position in a field, than the historical conjunction of particular kinds of ideologies, technologies, and actors. At a minimum, our cases, or the hundreds of others we might have chosen, represent entrepreneurial efforts to get others to be entrepreneurial, but not, it seems, purely for monetary gain or for status in a field of other purveyors of alternative diets.

It is not clear that the actors in this field are all struggling for power against each other, nor do their logics of action seem identical, despite common orientations to nutrition reform. Institutionalist theory, however, does not require that actors compete directly with each other, and instead it permits them to be uncertain, to muddle through, to find activities that fit what they know how to do whether they are winning or not, and to take seriously the idea that social movements can provide possibilities for changing conventions and rules for action in quite indirect ways, as in the case of the food counterculture and the rise of professionals influenced by the values of social movements. Institutions are not objective things in institutionalist theory, yet actors can be conscious of the rules and under particular circumstances can make concerted attempts to change them.

Given that science and technology have profound effects on the distribution of material and cultural goods and bads, it is not so much the exclusion of others from the easy-to-enter and loosely knit field of alternative nutrition that produces inequality. Inequalities are produced in this system by normalizing the updating of the body and emotional self, and via the translation of scientific ideas into commodities that are not expected to be lasting and satisfying, but require more purchases. Neoliberal entrepreneurialism, abetted by Internet technologies that change the speed of this kind of updating, normalize the instabilities that neoliberal economic policies have instantiated, particularly through the notion of risk self-management and the opportunity and information seeking that it requires. That the field through which this kind of activity is encouraged is ill defined may be precisely its power: as analysts of social networks have long understood, being at the intersections, rather than the center, can be very powerful indeed.

Yet the consequences for science more generally are significant, precisely because the flow of knowledge-as-commodity, its rapid production and distribution and its endless "updating" destabilizes traditional forms of authority in science. As the web – if not field – of alternatives to mainstream nutrition ideas proliferate, science itself is a weaker tool for battles over political and ethical questions about what to eat. This situation is not the result of explicit battles to weaken science, nor of battles for the center; the comingling of neoliberal economic logics of the enterprising self (Binkley, 2006, 2009) with metascientific analysis via charismatic authority, and the logic of American perfectionist impulses presents us with a situation in which fieldness itself is constantly being redrawn.

REFERENCES

Adams, J., Clemens, E. S., & Orloff, A. (Eds.). (2005). *Remaking modernity: Politics, history and sociology*. Durham, NC: Duke University Press.

Albert, M., & Kleinman, D. (2011). Bringing Bourdieu to science and technology studies. *Minerva*, *409*(3), 263–273.

Albert, M., Laberge, S., Hodges, B. D., Regehr, G., & Lingard, L. (2008). Biomedical scientists' perception of the social sciences in health research. *Social Science & Medicine*, *66*(12), 2520–2531.

Alkon, A. H., Block, D., Moore, K., Gillis, C., DiNuccio, N., & Chavez, N. (2013). Foodways of the urban poor. *Geoforum*, *48*, 126–135.

Ansell, C., & Vogel, D. (2006). *What's the beef? The contested governance of European food safety*. Cambridge, MA: MIT Press.

Baron, R. A., & Shane, S. A. (2008). *Entrepreneurship*. Mason, OH: South-Western.

Beecher, G. R., Stewart, K. K., Holden, J. M., Harnly, J. M., & Wolf, W. R. (2009). Legacy of Wilbur O. Atwater: Human nutrition research expansion at the USDA – Interagency development of food composition research. *Journal of Nutrition*, *139*(1), 178–184.

Belasco, W. (2006). *Appetite for change: How the counterculture took on the food industry* (2nd ed.). Ithaca, NY: Cornell University Press.

Bernstein, E. (2007). *Temporarily yours: Intimacy, authenticity, and the commerce of sex*. Chicago, IL: University of Chicago Press.

Binkley, S. (2006). The perilous freedoms of consumption: Toward a theory of the conduct of consumer conduct. *Journal for Cultural Research*, *10*(4), 343–362.

Binkley, S. (2007). *Getting loose: Lifestyle consumption in the 1970s*. Durham, NC: Duke University Press.

Binkley, S. (2009). The civilizing brand: Shifting shame thresholds and the dissemination of consumer lifestyles. *European Journal of Cultural Studies*, *12*(1), 21–39.

Binkley, S. & Capetillo, J. (Eds.). (2009). *A Foucault for the 21st century: Governmentality, biopolitics and discipline in the new millennium*. Newcastle Upon Tyne: Cambridge Scholars Publishing.

Bittman, M. (2008). *Food matters*. New York, NY: Simon and Schuster.

Bourdieu, P. (1975). The specificity of the scientific field and the social conditions of the progress of reason. *Social Science Information*, *14*(6), 19–47.

Bourdieu, P. (1993). In R. Johnson (Ed.), *The field of cultural production*. New York, NY: Columbia University Press.

Brandt, A. M. (1997). "Just say no": Risk, behavior, and disease in twentieth century America. In R. G. Walters (Ed.), *Scientific authority and twentieth century America* (pp. 82–98). Baltimore, MD: Johns Hopkins University Press.

Bromley, P., & Powell, W. W. (2012). From smoke and mirrors to walking the talk: Decoupling in the contemporary world. *Academy of Management Annals*, *6*, 1–48.

Clemens, E. S. (1997). *The people's lobby: Organizational innovation and the rise of interest group politics in the United States, 1890–1925*. Chicago, IL: University of Chicago Press.

Connell, R., & Dados, N. (2014). Where in the world does neoliberalism come from? *Theory and Society*, *43*, 117–138.

Cooper, K. H. (1968). *Aerobics*. Philadelphia, PA: M. Evans.

Department of Health and Human Services, & Department of Agriculture. (1977). *Nutrition and your health: Dietary guidelines for Americans*. Washington, DC: U.S. Government Printing Office.

Diamond, H., & Diamond, M. (1985). *Fit for life*. New York, NY: Warner Books.

DiMaggio, P. J. (1991). Constructing an organizational field as a professional project: The case of U.S. art museums. In W. W. Powell & P. J. DiMaggio (Eds.), *The new institutionalism in organizational analysis* (pp. 267–292). Chicago, IL: University of Chicago Press.

DiMaggio, P. J., & Powell, W. W. (Eds.). (1991). *The new institutionalism in organizational analysis*. Chicago, IL: University of Chicago Press.

Dorment, R. (n.d). Gwyneth Paltrow feels good – And so can you. *MSN Living*. Retrieved from (http://living.msn.com/life-inspired/gwyneth-paltrow-feels-good-andnumber8212-and-so-can-you). Accessed on July 1, 2013.

Dupont, J. (2009). Research in the agricultural research service. *Journal of Nutrition*, *139*(1), 171–172.

Eisinger, P. (1998). *Toward an end to hunger in America*. Washington, DC: Brookings Institute Press.

Elias, N. (1969). *The civilizing process. The history of manners* (Vol. 1). Oxford: Blackwell.

Emirbayer, M. (1997). Manifesto for a relational sociology. *American Journal of Sociology*, *103*(2), 281–317.

Espeland, W. N., & Stevens, M. (1998). Commensuration as a social process. *Annual Review of Sociology*, *24*, 312–343.

Fixx, J. (1977). *The complete book of running*. New York, NY: Random House.

Fletcher, C. (1971). *The complete walker*. New York, NY: Knopf.

Food and Nutrition Information Center. (n.d). *Home|food and nutrition information center*. Retrieved from http://fnic.nal.usda.gov/. Accessed on July 1, 2013.

Fourcade-Gourinchas, M., & Babb, S. (2002). The rebirth of the liberal creed: Paths to neoliberalism in four countries. *American Journal of Sociology*, *108*(3), 533–579.

Freedman, R., & Barnouin, K. (2005). *Skinny bitch*. Philadelphia, PA: Running Press Book Publishers.

Freedman, R., & Barnouin, K. (2009). *Skinny bastard*. Philadelphia, PA: Running Press Book Publishers.

Frickel, S., Campanella, R. J., & Vincent, B. (2009). Mapping knowledge investments in the aftermath of Hurricane Katrina. *Environmental Science and Policy*, *12*(19), 119–132.

Friedland, R., & Alford, R. R. (1991). Bringing society back in: Symbols, practices and institutional contradictions. In W. W. Powell & P. J. DiMaggio (Eds.), *The new institutionalism in organizational analysis* (pp. 232–266). Chicago, IL: University of Chicago Press.

General Accountability Office. (1978). *Nutrition research alternatives*. Washington, DC: U.S. Government Printing Office.

Go, J. (2008). Global fields and imperial forms: Field theory and the British and American empires. *Sociological Theory*, *26*(3), 201–228.

Goldman, M. (2012). *The American soul rush: Esalen and the rise of spiritual privilege*. New York, NY: New York University Press.

Graham, S. (2009[1839]). *Lectures on the science of human life*. Newcastle Upon Tyne: Cambridge Scholars Publishing.

Guthman, J. (2011). *Weighing in: Obesity, food justice, and the limits of capitalism*. Berkeley, CA: University of California Press.

Harvey, D. (1998). The body as an accumulation strategy. *Environment and Planning D: Society and Space*, *16*(4), 401–421.

Harvey, D. (2005). *A brief history of neoliberalism*. New York, NY: Oxford University Press.

Hess, D. J. (2007). *Alternative pathways in science and industry*. Cambridge, MA: MIT Press.

Hess, D. J. (2011). Bourdieu and science and technology studies: Toward a reflexive sociology. *Minerva*, *49*(3), 333–348.

Hess, D. J. (2012). A theory of fields: A review. *Mobilizing Ideas*, June 13. Retrieved from http://mobilizingideas.wordpress.com/2012/06/13/a-theory-of-fields-a-review/. Accessed on May 21, 2014.

Hess, D. J. (2014). Political ideology and the green-energy transition in the United States. In D. L. Kleinman & K. Moore (Eds.), *Handbook of science, technology and society*. New York, NY: Routledge.

Kotz, N. (1971). *Let them eat promises: The politics of hunger in America*. New York, NY: Doubleday Anchor.

Kripal, J. J. (2007). *Esalen: America and the religion of no religion*. Chicago, IL: University of Chicago Press.

Lappé, A. (2010). *Diet for a hot planet*. New York, NY: Bloomsbury USA.

Lappé, F. M. (1971). *Diet for a small planet*. New York, NY: Ballantine Books.

Lattin, D. (2010). *The Harvard Psychedelic Club*. New York, NY: Harper Collins.

Leschziner, V., & Green, A. I. (2013). Thinking about food and sex: Deliberate cognition in the routine practices of a field. *Sociological Theory*, *31*(2), 116–144.

Levenstein, H. (2003). *Revolution at the table: The transformation of the American diet*. Berkeley, CA: University of California Press.

Lounsbury, M., & Ventresca, M. J. (Eds.). (2002). "Social structure and organizations" revisited. *Research in the sociology of organizations* (Vol. 19, pp. 3–36). Oxford: JAI Press.

Maguire, J. S. (2008). *Fit for consumption: Sociology and the business of fitness*. New York, NY: Routledge.

Maney, A. (1989). *Still hungry after all these years: Food assistance policy from Kennedy to Reagan*. Westport, CT: Praeger Publishers.

Martin, J. L. (2003). What is field theory? *American Journal of Sociology*, *109*(1), 1–49.

Maynard, L. A. (1962). Early days of nutrition research in the United States of America. *Nutrition Abstracts and Reviews*, *32*, 345–355.

McAdam, D., & Fligstein, N. (2012). *A theory of fields*. New York, NY: Oxford University Press.

Metzl, J. M., & Kirkland, A. (Eds.). (2010). *Against health: How health became the new morality*. New York, NY: New York University Press.

Miller, J. W. (2005). Wellness: The history and development of a concept. *Spectrum Freizheit*, *27*(1), 84–106.

Miraval Resort. (n.d). Miraval Integrated Wellness Center. Retrieved from http://www.miravalresorts.com/the_experience/integrative_wellness_center/. Accessed on July 1, 2013.

Moore, K. (1996). Organizing integrity: American science and the creation of public interest science organizations, 1955–1975. *American Journal of Sociology*, *101*(6), 1592–1627.

Moore, K. (2008). *Disrupting science: Social movements, American scientists, and the politics of the military, 1945–1975*. Princeton, NJ: Princeton University Press.

Moore, K. (2013). Fear and fun: Science and gender, emotion and embodiment under neoliberalism. *The Scholar and Feminist Online*, 11, p. 1. Retrieved from http://sfonline.barnard.edu/gender-justice-and-neoliberal-transformations/fear-and-fun-science-and-gender-emotion-and-embodiment-under-neoliberalism/. Accessed on July 1, 2013.

Moore, K., & Hala, N. (2002). Organizing identity. *Research in the Sociology of Organizations*, *19*, 32–47.

Moore, K., Kleinman, D., Hess, D., & Frickel, S. (2011). Science and neoliberal globalization: A political sociological approach. *Theory & Society*, *40*(5), 505–532.

Nestle, M. (1993). Dietary advice for the 1990s: The political history of the food guide pyramid. *Caduceus*, *9*(3), 136–153.

Nestle, M. (2007). *Food politics: How the food industry influences nutrition and health* (Rev. ed.). Berkeley, CA: University of California Press.

Panofsky, A. (2011). Field analysis and interdisciplinary science: Scientific capital and exchange in behavioral genetics. *Minerva*, *49*(3), 295–316.

Penders, B. (2010). *The diversification of health: Politics of large-scale collaboration in health*. Beilefeld, DR: Verlag.

Pollan, M. (2007). *The omnivore's dilemma: A natural history of four meals*. New York, NY: Penguin.

Pollan, M. (2009). *In defense of food*. New York, NY: Penguin.

Polletta, F. (2009). How to tell a new story about battering. *Violence against Women*, *15*(12), 1490–1508.

Prevention. (1980). Now there's an owner's manual for the most valuable thing you own [your body]. *Prevention*, June, p. 101.

Prevention. (1981). Subtracting additives multiplies your health account. *Prevention*, January, p. 174.

Prevention. (1982). The surge to self-care. *Prevention*, December, p. 22.

Prevention. (1985a). Best and worst diet ideas. *Prevention*, January, p.22.

Prevention. (1985b). See if you need Vitamin C. *Prevention*, February, p.13.

Prevention. (1986). Eight reasons you might be going bald. *Prevention*, March, p.47.

Puttick, E. (2004). The human potential movement. In C. Partridge (Ed.), *New religions: A guide: New religious movements, sects and alternative spiritualities* (pp. 399–402). New York, NY: Oxford University Press.

Rejali, D. (2009). *Torture and democracy*. Princeton, NJ: Princeton University Press.

Robbins, J. (1987). *Diet for a new America*. Walpole, NH: Stillpoint Publishing.

Robertson, L., Flinders, C., & Godfrey, B. (1976). *Laurel's kitchen: A handbook of vegetarian cookery & nutrition*. Berkeley, CA: Nilgiri Press.

Rose, N. (2007). *The politics of life itself: Biomedicine, power and subjectivity in the twenty-first century*. Princeton, NJ: Princeton University Press.

Sarasvathan, S., Dew, N., & Ventresca, M. J. (2009). Unpacking entrepreneurship as collective activity: Opportunities, activity, and context. *Advances in Entrepreneurship, Firm Emergence and Growth*, *11*, 261–281.

Savage, M., & Silva, E. B. (2013). Field analysis and cultural sociology. *Cultural Sociology*, *7*(2), 111–126.

Schleifer, D. (2013). Categories count: Transfat labeling as a technique of corporate governance. *Social Studies of Science*, *43*(1), 54–77.

Schneiberg, M. (2012). Toward a pragmatist social science? Accomplishments and analytical challenges in the study of institutions. *Socio-Economic Review*, *10*(3), 594–600.

The Biggest Loser. (n.d.). *The biggest loser*. Retrieved from (www.biggestloser.com). Accessed on July 1, 2013.

Thelan, K. (2010). *Explaining institutional change*. Cambridge, MA: Cambridge University Press.

Thorton, P., Ocasio, W., & Lounsbury, M. (2012). *The institutional logics perspective: A new approach to culture, structure and process*. Oxford: Oxford University Press.

U.S. Senate Select Committee on Nutrition and Human Needs. (1977). *Dietary goals for the United States*. Washington, DC: U.S. Government Printing Office.

Venkatesh, S. (2013). Underground markets as fields in transition: Sex work in New York City. *Sociological Forum*, *28*(4), 682–699.

Weil, A. (1995). *Spontaneous healing: How to discover and embrace your body's natural ability to maintain and heal itself*. New York, NY: Random House.

Weil, A. (1997). *8 weeks to optimum health: A proven program for taking full advantage of your body's natural healing power*. New York, NY: Random House.

Weil, A. (1998[1972]). *The natural mind: A new way of looking at drugs and higher consciousness*. New York, NY: Mariner Books.

Weil, A. (2000). *Eating well for optimum health. The essential guide to food, diet, and nutrition*. New York, NY: Random House.

Weil, A. (2004[1983]). *Health and healing: Understanding conventional and alternative medicine*. New York, NY: Houghton Mifflin Company.

Weil, A. (2006). *Q&A library: Easy on the black pepper?* Retrieved from http://www.drweil.com/drw/u/QAA400102/Easy-on-the-Black-Pepper.html. Accessed on July 1, 2013.

Weil, A. (2007). *Healthy aging: A lifelong guide to your well-being*. New York, NY: Random House.

Weil, A. (2013a). *Fact sheet*. Retrieved from http://www.drweil.com/drw/u/PAG00070/Dr-Weil-Fact-Sheet.html. Accessed on July 1, 2013.

Weil, A. (2013b). *Anti-Inflammatory food pyramid*. Retrieved from http://www.drweil.com/drw/u/PAG00361/anti-inflammatory-food-pyramid.html. Accessed on July 1, 2013.

Weil, A., & Fox, S. (2012). *True food: Seasonal, sustainable, simple, and pure*. New York, NY: Little, Brown and Company.

Whittington, K. B., & Smith-Doerr, L. (2008). Women inventors in context. *Gender and Society*, *22*(2), 194–218.

Whorton, J. (2002). *Nature cures: The history of alternative medicine in America*. New York, NY: Oxford University Press.

Zinczenco, D., & Goulding, M. (2012). *Eat this, not that! 2013: The no-diet weight loss solution*. New York, NY: The Rodale Press.

Zolberg, A. (1972). Moments of madness. *Politics and Society*, *2*(2), 183–207.

Zoom Company Information. (2014). *Weil lifestyle, LLC*. Retrieved from http://www.lexisnexis.com. Accessed on April 13, 2014.

Zukin, S. (2008). Consuming authenticity: From outposts of difference to means of exclusion. *Cultural Studies*, *22*(5), 724–748.

WHAT IS VOLUNTEER WATER MONITORING GOOD FOR? FRACKING AND THE PLURAL LOGICS OF PARTICIPATORY SCIENCE

Abby Kinchy, Kirk Jalbert and Jessica Lyons

ABSTRACT

This paper responds to recent calls for deeper scrutiny of the institutional contexts of citizen science. In the last few years, at least two dozen civil society organizations in New York and Pennsylvania have begun monitoring the watershed impacts of unconventional natural gas drilling, also known as "fracking." This study examines the institutional logics that inform these citizen monitoring efforts and probes how relationships with academic science and the regulatory state affect the practices of citizen scientists. We find that the diverse practices of the organizations in the participatory water monitoring field are guided by logics of consciousness-raising, environmental policing, *and* science. *Organizations that initiate monitoring projects typically attempt to combine two or more of these logics as they develop new practices in response to macro-level*

Fields of Knowledge: Science, Politics and Publics in the Neoliberal Age
Political Power and Social Theory, Volume 27, 259–289
Copyright © 2014 by Emerald Group Publishing Limited
All rights of reproduction in any form reserved
ISSN: 0198-8719/doi:10.1108/S0198-871920140000027017

social and environmental changes. The dominant logic of the field remains unsettled, and many groups appear uncertain about whether and how their practices might have an influence. We conclude that the impacts of macro-level changes, such as the scientization of politics, the rise of neoliberal policy ideas, or even large-scale industrial transformations, are likely to be experienced in field-specific ways.

INTRODUCTION

Participatory environmental monitoring (PEM) – environmental monitoring carried out by non-scientist volunteers or activists – is now a widespread phenomenon in the United States. From the venerable Audubon Christmas Bird Count to the recent phenomenon of "hacker" or DIY water testing kits, diverse publics are engaging in work to monitor and measure changes to the environments in which they live. Much scholarship has celebrated the growth of this kind of public participation in science. For example, studies of volunteer water monitoring find that volunteers develop social capital and leadership skills (Lawrence, 2006; Overdevest, Orr, & Stepenuck, 2004; Stedman, Lee, Brasier, Weigle, & Higdon, 2009). Theorists concerned with expertise and democracy often use examples of PEM to demonstrate that laypeople are capable of contributing to debates that typically are restricted to scientific experts (Fischer, 2000; Kleinman, 2000).

However, recent studies suggest that the social implications of PEM are considerably more complex than the common narratives of "empowerment" and "democratization" suggest. Lave (2012), for instance, critically observes that "amateur and citizen scientists provide vast amounts of unpaid work for physical scientists," a symptom of broader neoliberal transformations in how science is conducted. Moore (2006) demonstrates that public participation can either undermine or reinforce the political authority of science, depending on what initiators establish as the purpose and practices of participation, suggesting the need for more comparative studies. Ottinger (2009, p. 248) argues that we should not be surprised by the "intermittent success" of citizen science because:

> A significant and obvious obstacle to citizen scientists' efforts to shape scientific policies and practices are the often extreme disparities of wealth, education, and power (among others) between them and those they seek to influence. As a result, understanding

> whether and under what circumstances citizen science can fulfill its transformative potential requires a deeper analysis, focused on these disparities, their institutional bases, and their consequences for competing knowledge claims.

This paper is a response to these calls for deeper scrutiny of the institutional contexts of citizen science. We seek to illuminate the processes of agenda-setting for participatory science that happen within fields – in this case, the field of organizations that have a stake in advancing participatory water monitoring projects in response to natural gas development in the northeastern United States. We focus on agenda-setting processes because the influence of participatory water monitoring projects is determined not just by power disparities, but also by project leaders who envision widely disparate paths by which their efforts might make a difference. By taking a field-level approach, our study differs from macro-level analyses that consider the broad social transformations associated with the growing prevalence of PEM in science and politics. Furthermore, by looking at multiple organizations interacting in a field, our approach also differs from many existing case studies of PEM, which typically focus on a single organization or project, rather than the larger organizational field in which those efforts occur.

Participatory (or "volunteer") water monitoring has a long history, but, in the states of Pennsylvania and New York, it has acquired new urgency and significance in the context of major natural gas drilling projects that began around 2008. Drilling companies use the controversial technique of hydraulic fracturing to extract gas from a geological formation called the Marcellus Shale. There are now at least two dozen civil society organizations that are monitoring the watershed impacts of the natural gas industry in those two states. These organizations make up a highly dynamic field, in which the traditional objectives of water monitoring projects – such as educating the public, keeping an eye on polluters, and providing data that is useful for scientific research – are being reevaluated and prioritized in new ways, due to new pressures related to shale gas extraction. We find that the diverse practices of the organizations in this field are grounded in three different "institutional logics," or organizing principles, that serve as a basis for action in the participatory watershed monitoring field. The first of these is what we refer to as a *logic of consciousness-raising*, which reflects democratic principles, as invoked by the broader field of environmental conservation organizations. The second, a *logic of environmental policing*, is closely related to the practices and expectations of the institution of environmental regulation. Finally, and crucially, the field is influenced by a *logic of science*. Organizations that initiate participatory water monitoring projects do not draw on each of these logics solely or equally. Many attempt to combine these logics as they develop new practices. We find that the dominant logic of the field remains unsettled, and many groups appear uncertain about whether and how their practices might have an influence.

In the next section of this paper, we situate this study in the literature about the prevalence of PEM in US science and politics and explain our theoretical framework, which draws on field theories in sociology. We then discuss our research methods. Next, we provide a brief history of the water monitoring field and begin to explain the three institutional logics that have shaped the practices of water monitoring organizations. We then describe how the threat of Marcellus Shale gas drilling catalyzed new water monitoring activities. This analysis considers the reasons for the growing emphasis on science and policing logics among project leaders and shows that some organizations are experimenting with mixed logics as they develop their agendas. We conclude by considering the broader implications of this study and discussing how field-level dynamics, as well as interactions between fields, shape the practices and agendas of participatory science.

LOGICS OF PEM

PEM has a long history in the United States, particularly in the form of amateur bird watching and weather observations; however, it grew rapidly in prevalence and importance after the 1970s (Lave, 2012; Moore, 2006). This shift is marked not only by the growing number of people involved in PEM, but also by its increasing relevance to policy debate and regulatory decision-making. Volunteer data gathering projects have in some cases been recognized as legitimate contributions to scientific research and regulatory processes. As Moore (2006, p. 304) points out, "participatory research is now a legitimated and fundable research method," supported by the National Institutes of Health and the Environmental Protection Agency (EPA). PEM practices are diverse, however, encompassing a wide variety of initiatives that differ in their goals and in the ways they position participants in relation to experts and other authorities (Moore, 2006; Shirk et al., 2012).

The growing prevalence of PEM suggests a changing relationship between science, citizens, and the state. However, scholars disagree about how to interpret these shifts. Some describe the scientization of society, which is due, in part, to the proliferation of invisible threats in the environment (Beck, 1992). In this view, the public is more involved in environmental monitoring because it is one of the only ways that they can gain knowledge of the problems they face and contest the claims of industrial opponents (Brown, 2007; Couch & Kroll-Smith, 2000; McCormick, 2006; Morello-Frosch, Brown, Altman, McCormick, & Mayer, 2006). Political sociologists of science have argued that PEM is part of larger process of epistemic modernization, in which the public scrutinizes science "from below," and, through social movement struggles, is increasingly involved in setting and pursuing research agendas (Hess, 2007; Moore, Kleinman, Hess, & Frickel, 2011). Finally, some scholars contend that PEM's visibility today is a result of neoliberal transformations in environmental policy, which, on one hand, has increased the exploitation of volunteer labor, but on the other hand has enhanced the credibility of "extramural" science (produced by industry or civil society organizations) (Lave, 2012).

Each of these perspectives offers valuable insights for interpreting the social significance of PEM today. However, PEM efforts are diverse, advancing multiple objectives. For instance, a recent study notes that desired impacts of PEM may include "sustained stewardship and conservation," "a knowledgeable and empowered community," and "responsive science" (Shirk et al., 2012). Others list a range of PEM goals, such as public education and scientific literacy, problem identification, provision of background information for local decision-making, habitat remediation, policy advocacy, and litigation (Savan, Morgan, & Gore, 2003, p. 562). In other words, while a general principle of supporting public participation in efforts to monitor the environment may be prevalent in many domains, the objectives and practices of these projects vary greatly and complicate their ability to be understood within a single theoretical framework.

Our approach, therefore, builds on Moore's (2006) insight that initiators (e.g., professionals, activists, or amateur scientists) have a key role in shaping the practices of participatory science, because "initiators set agendas, the terms of debate, and define the resources, languages, and venues for discussion and adjudication available." Professional scientists, amateur scientists, and activist groups are subject to different kinds of pressures when they initiate PEM projects. Professional-initiated PEM, such as a university-led research project that uses volunteers to collect data, might be understood as an appropriation of volunteer labor in the context of neoliberal budget cuts for basic science. In comparison, activist-led PEM, such as an effort to document pollution in a low-income community, might seem more closely associated with the bottom-up scrutiny of science described in the theory of epistemic modernization or the public response to the scientization of politics.

However, as Moore (2006) notes, PEM projects do not always have one initiator; often they emerge out of interaction between professionals and activists, between government agencies and amateurs, or other situations in which different fields interact. Therefore, to understand the mandates and practices of any single PEM project, one must analyze not just the initiators, but also the organizational field in which the agendas and practices of PEM efforts are established. We borrow Fligstein and McAdam's (2011, p. 3) definition of a strategic action field: "a meso-level social order where actors (who can be individual or collective) interact with knowledge of one another under a set of common understandings about the purposes of the field, the relationships in the field (including who has power and why), and the field's rules." In this case, we observe that PEM projects do not operate in isolation, but rather they are based in organizations that are part of a field in which "common understandings about the purposes of the field" may in fact be a subject of disagreement or in a process of change.

We understand these processes of change to stem from the contradictory relationships between institutions (Friedland & Alford, 1991), such as science, education, democracy, and the regulatory state. For example, Savan et al. (2003) describe their experience of hosting a citizen environmental monitoring initiative at a Canadian university, noting two pressures they sought to resist. They did not want academic pressures (such as the need to publish peer-reviewed work) to influence the goals and work of the grassroots groups they supported. At the same time, they resisted taking on the investigative and enforcement roles that they believed were rightfully the job of government, even though the government had effectively abandoned those responsibilities. The purpose of the citizen monitoring field, in that case, was defined in relation to competing pressures from academic and regulatory fields.

We find it particularly illuminating to examine the "institutional logics" that organizations in a field draw upon to respond to new challenges. An institutional logic is "a set of organizing principles for a major social order, such as the market, the state, the family, religion, or science. These principles explain the purpose of social action and serve as a basis for decisions about how to behave" (Berman, 2012, p. 9). Institutional logics provide rules and standards – guidance for action within a field.

The existence of institutional logics does not preclude substantial changes in practices over time. Logics are not iron-clad rules; rather, they are "available to organizations and individuals to elaborate" (Friedland & Alford, 1991, p. 248). Within an organizational field, participants may experiment with practices that are grounded in different logics; for example, university scientists increasingly engage in practices that are grounded in a market logic, such as entrepreneurship and patenting, despite the historical dominance of a science logic in the field of academic research (Berman, 2012). Individuals and groups can reinterpret symbols and practices and may transform one institution by importing the symbols and practices of another. Experimentation and change in the logics that are dominant in an organizational field is possible because of the multiplicity of institutions that make up the social order.

Changes in practices are often brought about by interactions between fields that draw on different institutional logics. Fligstein and McAdam (2011, p. 8) observe that all fields are "embedded in complex webs of other fields." This means that participants in one field (such as academic science) may come to adopt logics from proximate fields (like the market), through interchanges between actors in the fields. In an emerging field, different institutional logics may come into conflict, or combine, when organizations from different fields, such as universities, environmental advocacy groups, regulatory agencies, or community associations, interact. Interdependence among fields also means that changes in one field may bring about disruptions in other fields. "[T]he interdependence of fields is … a source of a certain level of rolling turbulence in modern society. A significant change in any given [strategic action field] is like a stone thrown in a still pond, sending ripples outward to all proximate fields" (Fligstein & McAdam, 2011, pp. 8–9).

Often, multiple logics persist over time in a field, with no clearly dominant logic. For example, the field of medical education in the United States has been guided by the plural logics of science and care since at least 1910 (Dunn & Jones, 2010). However, the relative dominance of each of these logics has fluctuated over time as a result of competition with rival fields, such as public health, conflicts among physicians within the field, and the changing demographics of medical professionals (Dunn & Jones, 2010).

In the analysis that follows, we examine the logics that inform organizations' use of PEM as a response to the new threat of shale gas development. In particular, we ask how these practices are affected by the organizations' relationships with the fields of university science and the regulatory state. Changes in the funding structure and practices of academic science have had ripple effects on the field of civil society organizations that promote volunteer water monitoring. Academic scientists increasingly depend on "citizen scientists" to gather data for them. This relationship provides some valuable resources to the field of PEM, and it also imports a science logic into the field that may come into conflict with preexisting logics of consciousness-raising and public engagement in environmental problem-solving. Standards and rules from the scientific field introduce new practices – and constraints on practice – to the participatory water monitoring field. At the same time, the entrance of a new industry that strains the capacity of regulators also creates turbulence across fields, in this case leading PEM organizations increasingly to take on practices that reflect a logic of law enforcement, or what we call an environmental policing logic.

METHODS

Data for this study includes interviews, field notes, and a survey of civil society organizations that are involved in surface water monitoring and watershed protection efforts in New York and Pennsylvania. The purpose of the survey was to identify organizations that are monitoring surface water for impacts of Marcellus Shale development and to gain some understanding of the technical and social dimensions of these efforts. We contacted 219 civil society organizations, identified from public listings of organizations that are involved in watershed protection and monitoring. We received responses from 188 organizations, 76 of which were actively monitoring watersheds. Of those, 24 reported that they were specifically monitoring the impacts of Marcellus Shale gas development. This paper focuses on those 24 organizations, as well as other organizations, scientists, and government agencies that are influential in shaping the field. The survey provides insight into their goals, practices, and collaborations. Results from the survey were compared to earlier published studies of the broader participatory water monitoring field in the United States and in the state of Pennsylvania, in order to identify continuities and unique qualities of the organizations monitoring shale gas development.

In addition, from 2011 to 2013, we were participant observers in an array of meetings, trainings, and conferences for participatory water monitoring organizations. These include volunteer trainings and outreach events hosted by six different organizations, as well as several conferences in which multiple water monitoring organizations were engaged in discussion and exchange of ideas. Field notes from these events provide another source of data for this analysis. Additionally, we interviewed representatives of nine civil society organizations in the water monitoring field and engaged in follow-up correspondence with those informants. Interview transcripts were coded using qualitative data analysis software (Atlas.ti) for concepts related to the priorities and practices of participatory monitoring efforts, the nature of interactions among organizations in the field, and the influence of academic and regulatory science. From these coded quotations, patterns were identified and comparisons were made (1) between organizations and (2) with survey results and field notes.

PARTICIPATORY WATER MONITORING

Participatory water monitoring has a long history in the United States. In 1926, the Izaak Walton League of America (IWLA), a group of anglers who came together to protect the health of rivers, organized its first survey of water quality problems in the United States (Firehock & West, 1995). The League eventually initiated the Save Our Streams (SOS) program, one of the earliest efforts to systematically train volunteers to monitor streams. The program aimed primarily to increase public awareness of watershed issues. One of that program's leaders wrote in 1995, one of that program's leaders wrote, "The primary goal of SOS is to motivate people to change behavior and to get people involved in preventing pollution and restoring streams. Volunteer monitoring is a tool to achieve that goal" (Firehock & West, 1995).

Not only conservation groups like the IWLA are not the only ones to have initiated participatory water monitoring projects. Throughout the 1980s and 1990s, water quality specialists in government debated whether existing systems for monitoring water quality were producing useful knowledge (for an overview of this debate, see Ward, 1996). By the end of the 1980s, the EPA was beginning to refer to citizen "stream watch" initiatives among other possible strategies to improve the nation's water monitoring capacity. This occurred at a time when the EPA and other environmental agencies were facing severe budget cuts. In the 1980s, many environmental initiatives were devolved to the local levels, where "volunteer inputs [were] very attractive" (Pfeffer & Wagenet, 2007, p. 237). Some states were using volunteers to collect water quality data needed for management purposes, although, as Nerbonne and Nelson (2004) found, few states actually integrated volunteer-collected data into regulatory and management programs. In a 1990 guidance document for states, the US EPA identified these volunteer water monitoring programs as possible sources of water quality knowledge that could satisfy public demands and perhaps provide a low-cost source of data to cash-strapped agencies (see, e.g., US Environmental Protection Agency, 1990). Federal and state government agencies began actively promoting and supporting volunteer stream monitoring. By 1992, at least 32 states had citizen monitoring programs (Pfeffer & Wagenet, 2007, p. 239). The EPA continues to sponsor national conferences for volunteer organizers, publishes manuals on volunteer monitoring methods, and maintains a searchable (though rather outdated) database of volunteer monitoring organizations.

The two states considered in this paper have had moderate involvement in these types of projects. New York boasts a long-standing Citizens Statewide Lake Assessment Program, in which volunteers gather data to support the management of lakes and ponds across the state.[1] In Pennsylvania, the Citizens' Volunteer Monitoring Program (CVMP) provided training, equipment, and administrative support to 11,000 volunteers in 138 watershed groups across the state (Wilson, 2002), until being disbanded in 2009 due to budget cuts. Pennsylvania's Department of Environmental Protection (DEP) says it will use data collected by volunteers if it meets a set of quality assurance and quality control (QA/QC) guidelines published by the agency – although few organizations have been able to meet these standards.[2] Neither of these state-run programs is connected to the volunteer monitoring projects that are focused on the impacts of the gas industry.

The participatory water monitoring field in the United States has historically operated with at least three different objectives – educating the public about conservation issues, producing scientific knowledge, and policing violations of environmental law. The first objective reflects a logic of consciousness-raising, which we believe stems from a widely-held idea that in a democracy, social change occurs through education of the citizens. Organizations guided by this logic pursue a strategy of environmental protection through increased public understanding and education about natural resources (Pfeffer & Wagenet, 2007, p. 239). The SOS project of the IWLA, described above, is a good example of this approach. In 2004, a national study of volunteer stream monitoring organizations found that these groups overwhelmingly ranked "public education" as their most important goal (Nerbonne & Nelson, 2004). In contrast, "only one-third of volunteer groups identified enforcement [of environmental law] as high priority, while improving legislation appeared to be the lowest priority, with one-fifth of the responders considering it 'not a goal'" (Nerbonne & Nelson, 2004, p. 827).

Some see the long-standing logic of consciousness-raising as in tension with a science logic, which prioritizes systematic and standardized data collection that can be used for hypothesis testing or tracking trends over time. For instance, a founder of the SOS program stated, "It is a shame that some programs have become more of a monitoring tool ... generating numbers and trend analyses ... rather than serving as a tool for people to see the presence of problems and learn what they can do" (Richard Klein, as quoted in Firehock & West, 1995, p. 201). The trend that Klein laments is likely due to the growing interest of government agencies and scientific researchers in using volunteers, rather than paid staff, to gather water quality data. In addition, many academic scientists have taken an interest in participatory models of environmental research (Shirk et al., 2012) and some have sought to use "citizen science" to facilitate data collection for large research projects. Since the 1990s, there have been numerous studies seeking to establish whether volunteers can "do real research" (Cohn, 2008) or serve as an "ecological research tool" (Dickinson, Zuckerberg, & Bonter, 2010). Participatory water monitoring projects initiated by academic scientists to accomplish research tasks reflect the logic of science, though it is often combined with a consciousness-raising logic (e.g., in attempts to promote scientific literacy among volunteers).

Finally, some participatory water monitoring projects are guided by what we to "call an environmental policing logic, which emphasizes observation of regulatory violations and reporting of environmental problems to regulatory authorities. Overdevest and colleagues (2004, p. 177) use the example of "observers who monitor for Clean Water Act violations on industrial timber harvests in the southern U.S." as an illustration of this type of monitoring. Although these practices are not often cited as a primary objective for organizations in the water monitoring field, there are numerous examples of volunteer monitors in other fields (such as air quality monitoring) drawing attention to regulatory violations (O'Rourke & Macey, 2003; Ottinger, 2009). Some researchers have drawn explicit comparisons between PEM and the practices of community policing, such as neighborhood watch efforts in high-crime areas (O'Rourke & Macey, 2003). Lynch and Stretesky (2013), in particular, compare citizen-led water monitoring efforts to community-oriented policing. Surveys of organizations in the water monitoring field suggest that environmental policing has been a subordinate logic in field when compared to the logic of consciousness-raising (Nerbonne & Nelson, 2004, 2008). However, we find that policing is beginning to take a more central role in the response to the threat of Marcellus Shale development.

MONITORING GAS DEVELOPMENT

In Pennsylvania and New York, the water monitoring field is undergoing a transformation, spurred by the environmental threat of shale gas development. The United States is in the midst of a natural gas boom, facilitated by new hydraulic fracturing drilling techniques, often referred to as "fracking," that make it possible to extract gas from very deep and dense shale formations. In the last decade, oil and gas companies have flocked to the Marcellus Shale, a geological formation stretching across several northeastern states. The arrival of the gas industry brings a web of drilling pads, pipelines, compressor stations, wastewater facilities, and other infrastructure. These operations create many possible pathways for water, air, and soil pollution, with potentially significant public health and ecosystem consequences. Gas extraction may impact watersheds in a variety of ways (Entrekin, Evans-White, Johnson, & Hagenbuch, 2011; Rahm & Riha, 2012; Soeder & Kappel, 2009). Water withdrawals may reduce the quantity of water in streams and other bodies of water, affecting aquatic habitats. Pollution may result from accidental or intentional releases of salty, possibly hazardous wastewater into soil and streams. Spills of toxic chemicals used in drilling could pollute surface water, while underground migration of methane and drilling fluids could potentially affect aquifers. Additionally, there is the likelihood of increased storm runoff due to the clearing of forests for pipelines, well pads, and other infrastructure, as well as the construction of new roads. Some negative impacts of Marcellus Shale development have already been discovered in Pennsylvania watersheds (States et al., 2011; The Academy of Natural Sciences of Drexel University, n.d.).

Concerns about shale gas development prompted several organizations in the water monitoring field to develop new monitoring protocols related to gas impacts and to train new volunteers in the techniques. Often, these organizations are motivated by a desire to maintain the water quality that they have worked hard to attain. As one long-time watershed activist told us, "as we learned more [about shale gas], we … realized, oh my gosh, all that we have been doing for all these years, working to protect the [watershed] and working on proactive efforts to get special protection waters in place …. All of those could be really undone by this industrial activity and that we had to really get ahead of it" (5-21-12 2).[3] Several well-established organizations developed projects to monitor the impacts of gas development, including the Delaware Riverkeeper Network, working in the Delaware River Basin, and Mountain Watershed Association (MWA), working in the Ohio River Basin, each of which has a long history of working with volunteers to identify and solve problems in their respective watersheds. In many cases, existing watershed associations – community organizations created in decades past to address watershed problems such as soil erosion, agricultural runoff, or acid mine drainage from the coal industry – have started collaborating with local government or with other organizations to begin monitoring the impacts of the gas industry. Participatory water monitoring efforts in the Marcellus Shale region are funded by a number of local and regional nonprofit organizations such as the Colcom Foundation and the Western Pennsylvania Conservancy.

The tools used to monitor the impacts of the gas industry are fairly similar across organizations, but the organizations vary widely in their missions and strategies for protecting watershed health. One of the earliest efforts was launched by the Pine Creek Headwaters Protection Group, with guidance from the Tioga County Planner in north central Pennsylvania. In 2009, soon after shale gas development began in Pennsylvania, they began training volunteers to be watchdogs of the natural gas industry – or "Waterdogs," as these volunteers came to be known. The monitoring protocols include using a hand-held device to measure the conductivity of surface water – an indicator of salts and metals – and observing disturbances to land and water. At their formation, the Waterdogs drew on an environmental policing logic. Organizers typically described volunteers as extra "eyes and ears" for regulators, to ensure that problems did not go unnoticed.

A second early initiator of shale impacts monitoring projects is the Alliance for Aquatic Resource Monitoring (ALLARM) at Dickinson College in Carlisle, Pennsylvania. ALLARM has a long history within the Pennsylvania water monitoring community since it began offering training support in 1986 to track the impacts of acid rain. At the request of their affiliated community groups in 2009, at the request of their affiliated community groups, ALLARM began to develop a new set of monitoring practices to respond to Marcellus Shale development. Groups trained by ALLARM use monitoring protocols vetted by regulatory agencies with which ALLARM has developed long-standing relationships over the years. Since 2010, ALLARM has conducted 22 training sessions throughout Pennsylvania.[4] The ALLARM monitoring protocol has become a key reference for other water monitoring organizations. ALLARM combines democratic consciousness-raising with scientific practices, as described on that organization's website:

> ALLARM enhances local action for the protection and restoration of Pennsylvania watersheds by empowering communities with scientific knowledge and tools to implement watershed assessments. Through the work of student and professional staff, ALLARM offers comprehensive services to enable groups to use critical scientific tools to enhance environmental quality and fully participate in community decision-making.[5]

A third initiator of new monitoring projects is the Pennsylvania Council of Trout Unlimited (PATU), a state-level branch of the nation-wide Trout Unlimited advocacy organization. The primary mission of PATU is to serve their member base of sporting enthusiasts by protecting high-quality fishing habitats in thousands of minor tributaries of Pennsylvania's six large watersheds. PATU oversees 50 local chapters and 12,000 members across the state. A number of chapters began monitoring the watershed impacts of gas development in 2010, as part of their Coldwater Conservation Corps (CCC).[6] The CCC builds on the training resources of ALLARM, but also has hired full-time staff and modified the protocols to align with the organizational mission to maintain trout populations. Volunteers "focus on achieving early detection of pollution events during oil and gas drilling and production activities and collecting a baseline inventory of data on important coldwater fisheries."[7] The emphasis of the project is to collect data that may reveal long-term, cumulative impacts of shale gas development across the state.

The Community Science Institute (CSI), in Ithaca, NY, illustrates a fourth type of initiator of participatory water monitoring. CSI is an independent nonprofit founded in 2000 by a PhD scientist to encourage and support volunteer watershed monitoring, backed by water testing in CSI's certified laboratory. Like ALLARM in Pennsylvania, CSI developed a protocol for monitoring the impacts of gas development at the prompting of a concerned landowner. In this case, the initiative was to enable communities throughout the southern tier of New York to gather baseline water quality measurements before the onset of gas development. CSI's "red flag" monitoring program involves rigorous volunteer training in a monitoring protocol that is somewhat more complex than those of other initiators. CSI staff scientists conduct laboratory tests to check the accuracy of volunteer measurements, maintain a database, and report on trends found in the data. While clearly guided by a logic of science, CSI emphasizes its difference from academic science, focusing on "community science" that addresses "local issues and local government" and uses "results to manage local resources sustainably."[8]

As these examples suggest, many of the initiators of gas-related water monitoring projects have extensive experience with participatory stream monitoring, predating the threat of fracking. Many describe their goals and practices in ways that reflect the logic of consciousness-raising that has long been dominant in the broader participatory water monitoring field. However, the threat of shale gas development has given greater prominence to the logics of science and policing, for multiple reasons.

First, concerns about the impacts of gas development have prompted collaboration between organizations that previously approached watershed protection with very different strategies. Organizations that do policy advocacy and environmental litigation are now networked with organizations that have historically used non-confrontational strategies to promote conservation. Initiators have begun to collaborate and offer trainings to numerous organizations and independent volunteers. For example, ALLARM has offered trainings to groups of volunteers organized by Trout Unlimited and MWA, among others. This has led to an exchange of ideas about the agendas of water monitoring projects.

Second, as university scientists have increasingly taken an interest in studying the impacts of gas development on water quality, a science logic is gaining greater prominence in the field. Researchers at several colleges and universities in Pennsylvania and New York are now collaborating with volunteer groups to gather water quality data related to gas drilling. University-based initiatives to compile volunteer-collected data in online databases have further linked these organizations together, which we discuss below.

Finally, as the failures of government agencies to monitor and enforce regulatory violations have become more evident, many organizations have begun to adopt a policing logic, in addition to their long-standing commitments to public education and scientific knowledge production. It is widely believed that Pennsylvania's environmental regulatory agencies are understaffed and overwhelmed by the volume of permitting and enforcement issues arising in conjunction with the development of shale gas. Many monitoring organizations cite insufficient government oversight of the natural gas industry as a key reason for volunteer monitoring.

Over time, and through collaboration, organizations in the field have shifted their agendas and practices to reflect new combinations of institutional logics. Many organizations in the shale impacts monitoring field express multiple objectives, reflecting the prevalence of mixed logics of consciousness-raising, science, and environmental policing. We address each of these logics separately, noting examples in which organizations combine them or find that they are in conflict.

Raising Awareness

Our survey of 24 organizations that are monitoring the impacts of Marcellus Shale development demonstrates the continuing prevalence of logic of consciousness-raising in this field. We asked organizational leaders to identify their groups' objectives from a list ("check all that apply"). Nineteen reported that they aimed to "inform and educate the public" and 14 said they wanted to "strengthen relationships in the community," among other objectives. Fewer than half (42%) of respondents selected one or more of the following objectives: "improve regulation of the natural gas industry," "change industry behavior," or "support litigation." When compared to other studies (Nerbonne & Nelson, 2004; Stedman et al., 2009), this finding suggests that interest in policy or regulatory outcomes is higher among these organizations than in the water monitoring field as a whole. Nevertheless, the majority of organizations in this field are guided by a logic of consciousness-raising, rather than explicit political advocacy goals.

When asked about their organizational missions, survey respondents that did not report political advocacy goals typically referred to promoting "awareness," "stewardship," and "conservation," partnering with schools to teach young people about watershed issues, "bringing people together," and "connecting people to nature." Some organizations in the water monitoring field that do not do water monitoring themselves, but train and support monitoring groups, also emphasize consciousness-raising. For example, the Water Resources Education Network (WREN), a project of the League of Women Voters of Pennsylvania, has funded a variety of efforts to monitor the impacts of shale gas development. WREN lists this as its first goal: "To foster and support local stakeholder communities which will educate themselves, other citizens, and local officials about their water resources and the public policies necessary to protect them."[9] While policy outcomes are mentioned, the means of achieving them is through environmental education.

One of the main ways that water monitoring groups have sought to make watershed information available to the public is through constructing websites and databases. One leader of a watershed association referred to the educational value of monitoring shale impacts on water, describing a partnership with a group of academic scientists that would enable them to share water quality information with the public:

> A database is being developed that will house data consisting of water quality parameters, macroinvertebrates, electrofishing surveys,[10] etc. This allows the public to click on a blip and it will give you all of those details for a particular stream reach. So fisherman can see what bugs are there at different times, locations that trout and other fish species are holding and the overall health (chemistry) of the stream. Similar monitoring projects are located throughout the rest of the shale basin. (11-15-12 1)

This quotation suggests that even as university scientists are increasing their involvement in civil society efforts to monitor the impacts of fracking, informing the public about water quality has remained a priority. Building databases requires technical expertise, often found within universities. Academic researchers are interested in building databases of volunteer monitoring data to support their research aims, but watershed groups have a different goal: to enable concerned citizens to access information about water quality that otherwise would not be available to them. The construction of water quality databases that successfully incorporate both scientific and consciousness-raising logics has been an ongoing challenge and source of tension within the water monitoring field, as discussed below.

Many organizations are successful at combining multiple logics. The leader of one volunteer monitoring organization noted that consciousness-raising practices can generate data that is useful for scientific research:

> [Our mission] is to empower citizens to understand and protect and manage their natural resources and in our case, the focus is on water. So that is the main value, and in the process of empowering volunteers and educating volunteers, you know, you collect a lot of very, very good data, because the volunteers are partnering with a certified lab. So all the data is really usable by any government entity or academic entity that cares to use it. (4-11-12 2)

In another example, MWA describes a consciousness-raising approach on its website, along with local action for conservation: "Our major purposes include … developing community awareness, promoting cooperative community efforts for remediation and encouraging sound environmental practices." MWA's community awareness mission is evident in their coordinating volunteer monitoring efforts in partnership with ALLARM through their "Marcellus Citizen Stewardship Project."[11] However, like many other organizations in the shale impacts monitoring field, MWA is now using practices of data collection and analysis that rely on expensive equipment and trained technical staff. In keeping with regulatory standards and a policing logic, it has a protocol for reporting suspected violations of environmental law to regulators.[12] Further examples in which plural logics are shaping practices of shale impacts monitoring are discussed in the next two sections.

Building Scientific Datasets

Of the 24 organizations surveyed that are conducting shale impacts monitoring, 20 indicated that one of their objectives is to contribute to scientific knowledge. A science logic has long been present in the participatory water monitoring field, as both government programs and academic scientists have turned to volunteers to gather data for them. However, facing the new threat of shale gas drilling, academic scientists have acted quickly to develop relationships with water monitoring organizations and to incorporate volunteer observations into useable datasets. As a result, a science logic is gaining greater prominence in the field. Several new practices are notable in this regard: the adoption of standardized QA protocols, the use of automated data loggers (sensing devices) to gather data rather than relying on volunteers, and the construction of large databases housed at universities.

There are clear reasons why academic researchers are interested in using data collected by volunteers to analyze the impacts of shale gas development. There is little known about how gas development might impact watersheds. Environmental scientist David Velinsky, for example, has pointed out that the cumulative impacts of gas development are poorly understood. It is not known:

> [W]hether there is a threshold point past which a certain density of drilling activity has an impact on the ecological health and services of the watershed regardless of how carefully drilling is conducted. Past studies that have looked at particular well sites or particular incidents fail to give a picture of the chronic impacts that might be expected from drilling and especially hydraulic fracturing. (Velinsky, 2010)

Yet, water quality data to answer such questions is in short supply. Government water monitoring programs are the primary source of information about the impacts of shale gas development, but these are limited. Therefore, some academic scientists have looked to volunteers as a source of data that might answer their questions.

An important new development is the construction of online databases for water quality data, hosted at universities. One of these is the Three Rivers Quest (3RQ) project, based at West Virginia University (WVU), which has brought funding and computing resources to organizations monitoring in southwestern Pennsylvania. The 3RQ strategy is to work with three "research partners" in different regions of western Pennsylvania: the Iron Furnace Chapter of Trout Unlimited, Duquesne University, and Wheeling Jesuit University. Each of these partners collects biweekly water samples for a long-term WVU study of many water quality indicators. But each is also responsible for offering mini-grants to facilitate monitoring projects by watershed associations and volunteer groups in their respective regions.[13] Data from all of these participatory monitoring projects is being incorporated into the 3RQ database.

Another database project is called Shale Network. In 2011, a group of researchers at Pennsylvania State University (Penn State) received a National Science Foundation (NSF) grant to compile and analyze volunteer-generated data about the watershed impacts of Marcellus Shale gas development. Shale Network describes itself as an "'honest broker' that collates datasets and learns and teaches how to synthesize that data into useful knowledge."[14] Meetings organized by Shale Network have brought together representatives of many organizations in the field to exchange ideas about data sharing. According to geoscientist Susan Brantley, the principal investigator on the project, "In the future, many monitoring networks of all kinds will need to include citizen scientists to keep costs down, and research scientists will need to learn to use such networks to [achieve] the best outcome" (National Science Foundation, 2011).

This statement, quoted in a profile on the NSF website, suggests the kind of "appropriation" of volunteer labor that Lave (2012) describes in the context of neoliberal changes in the practices of science. Unsurprisingly, not all organizations in the shale impacts monitoring field are supportive of the model of citizen science that academic researchers tend to pursue. For instance, the head of one organization criticized what is typically known as "citizen science":

> I think of citizen science as basically the standard big science model, where you have researcher at a university who gets a grant and they want to study birds in North America. Well, they don't have enough graduate students to send them all over North America, so what do they do? They invent citizen science and now they can get people to record data for them, for free. But, the scientists have complete control; there is very little input from the volunteers. The volunteers don't get to choose, necessarily, very much of anything about the project. So the project is largely controlled by the scientists and the output from citizen science is a peer reviewed publication, which may or may not impact policy. (4-11-12 2)

Academic researchers' search for low-cost methods of collecting data, and the increasing coordination among participants in these projects, is playing an important role in shifting the character of monitoring for shale gas impacts. Scientists' expectations for collaborations with volunteers in some cases generate practices that come into conflict with previous logics of consciousness-raising and practices of public engagement in environmental problem-solving. This was evident in the critique of university-led "citizen science," quoted above. But it is also evident in the skepticism that some local monitoring groups express about the efforts to synthesize data in databases that are intended to facilitate scientific analysis. In a discussion at a training session, one volunteer water monitor said, "My fear is that these groups are going to get a big grant and build the databases. Then they don't maintain it. But down the road, if they don't work out, we have to chase down our data with someone else. We need to keep that data local for our use, most importantly" (field notes, volunteer training follow-up session, central Pennsylvania, November 16, 2012).

Civil society organizations in the field are typically aware of the scientific standards that academic and regulatory scientists use in their analysis of water quality, and many seek to align their own monitoring practices with those of recognized experts in the field. For some organizations, this means processing volunteer-collected samples in state certified testing laboratories and, for many, it means following quality QA/QC procedures that are perceived to strengthen the credibility of their observations. Such practices were not unheard of in the participatory water monitoring field prior to fracking; indeed, any organization providing data to state agencies would be expected to follow strict QA/QC procedures and some organizations, like CSI, have used a certified lab for many years.

However, concern for QA/QC appears to be heightened in relation to fracking. Of the 24 survey respondents, one-half said that they followed QA/QC procedures. One leader told volunteer trainees, "The reason our data is getting attention and respect is because of our QA/QC. For example, you are all sitting here for 6 hours, calibrating your equipment regularly, doing split samples, sending replicates to the lab" (field notes, volunteer training session, central New York, June 16, 2012). Most common QA/QC methods include regular calibration of monitoring tools, duplicating field measurements, and sending "split samples" to a laboratory to check the accuracy of their field calculations. Many organizations compare and evaluate the strength of their own QA/QC practices to those of others in the field. In at least one instance, a water monitoring organization has sought to protect the credibility of their data by avoiding partnerships with organizations that have seemingly less-rigorous research methods (field notes, regional watershed conference, central Pennsylvania, May 20, 2013).

Some monitoring organizations are also adopting expensive technologies to produce enormous datasets of quantitative water quality measurements, a move that breaks with past practices of raising awareness through hands-on volunteer monitoring experiences. With the support of local foundations, area universities, and county conservation districts, several organizations across Pennsylvania have begun to use continuous data loggers to gather water quality data. These devices record measurements for weeks at a time, which are then downloaded from the devices by organization staff or sometimes volunteers.

Of the 24 monitoring organizations in our survey, five operate continuous data loggers for tracking the impacts of natural gas extraction. Many think these devices are more dependable than volunteer water monitors and have greater scientific credibility because they remove the possibility of human error or bias. Generally, analysis of measurements from data loggers is done by professional staff, not volunteers, because it requires a high level of technical skill. However, some organizations are using both volunteers and data loggers, and one organization uses continuous data loggers to check the accuracy of measurements made by volunteers, explaining, "One of the goals [is] to combine the data logger program with the volunteer monitoring program and get a volunteer monitor on the same page as data loggers, so that if a volunteer gets a peak, then we can go download the data logger and have more quality control on what is going on in that stream" (8-2-12 1).

Nearly all of the data logger programs are supported by funds from Colcom, a Pittsburgh based organization that "favors programs that aggressively address watershed remediation, natural resource preservation, clean air and improved water quality, and farmland and wildlife habitat conservation."[15] One organization received more than $25,000 to support their data logger network.[16] In another instance, the Colcom Foundation and another regional foundation provided funds to enable a coalition of watershed associations to install more than 50 data loggers.[17] Data loggers are also explicitly tied to academic databases being built by academic institutions. Colcom, for example, one of Colcom's primary objectives in funding 3RQ is to find a home for data generated by data loggers used in WVU's scientific water quality studies.

In addition to the integration with university-based projects, some data logger networks are explicitly designed to dovetail with government monitoring programs. One organizer told us about her organization's recent collaboration with the Army Corps of Engineers, which she saw as an important form of recognition for their work.

> The [river we focus on] is managed by the Army Corps of Engineers, so we are about to build a partnership with them to have data loggers on their site, and they actually paid [for the] data loggers and … contract [with our staff] to maintain them and record the data to share with them. So it's the first relationship with the Army Corps of Engineers with just, like, a non-profit citizens group. So to have that was a big stepping stone, to have a federal agency be like, yeah, let's work together on this. (8-2-12 1)

In several other examples, watershed associations are collaborating with county government agencies, which often house and analyze the data produced by the sensing devices.

In summary, participatory stream monitoring is seen by civil society organizations and scientists alike as a source of valuable scientific data about the impacts of shale gas development – if procedures are in place to ensure that the data is of good quality. The emphasis on QA/QC procedures in the shale impacts monitoring field reflects the influence of the academic and regulatory science fields, both before and since the initiation of shale gas development. The prevalence of a science logic is also evident in the increasing use of databases – specifically in response to fracking – that not only enable volunteers to track and visualize their own data, but also compile data in ways that are useful for academic partners in testing hypotheses and analyzing water quality trends over time. Furthermore, the growing use of continuous data loggers, rather than volunteers, to collect water quality data suggests alignment of the practices of professional scientists and water monitoring organizations. The institutional logic of science, however, is not the only logic that is rising in prominence in the shale impacts monitoring field. Organizations in the field are also increasingly adopting practices that reflect a logic of environmental policing.

Environmental Policing

The growth of shale gas development has strained regulatory agencies, and this has had ripple effects, leading PEM organizations increasingly to take on environmental law enforcement practices, which we call "environmental policing." Of the 24 survey respondents, 9 listed "change industry behavior," 6 indicated "support litigation," and 10 listed "improve regulation of the natural gas industry." These responses suggest that fewer than half of the organizations monitoring the impacts of fracking explicitly see themselves as operating with a policing logic. Nevertheless, even if they did not report those objectives, nearly all of the organizations said they planned to report observations of pollution or other damage caused by gas development to environmental regulators. In many ways, the practices of these organizations resemble "neighborhood watch" efforts that rely on volunteers to observe and report suspicious activity.

Water pollution stemming from environmental regulatory violations is a matter of serious concern to communities that depend on local fresh water sources, whether for drinking supplies, watering livestock, irrigating fields, fishing, or recreation. Thus, many shale impacts monitoring organizations see their efforts as extending regulatory agencies' "eyes and ears on the ground" (a phrase frequently repeated among participants in the field), and

15 of the 24 organizations have protocols in place for reporting observations of pollution or other disturbances to regulators. One Pennsylvania university scientist who works with volunteers shared the perception that volunteers are needed to extend the capacity of regulators:

> Recognizing that governmental agencies charged with environmental oversight have limited manpower and resources, individuals and watchdog groups are stepping forward to take action. Not only do these additional eyes and ears on the ground provide important field surveillance assistance, but individuals trained to use basic monitoring equipment can generate valuable preliminary water-quality data.[18]

Government agencies are not necessarily using volunteer-collected data. Indeed, our interviews with regulatory officials indicated that they could not use volunteer data for enforcement purposes, although they might follow up on a complaint by sending their own staff into the field. However, one watershed association leader expressed confidence that both volunteers and data loggers have an important role in supporting regulatory enforcement. In an email, he explained:

> With this recorded data, we have an exact time frame of the event (start to end) and we have recorded the fluctuation of various water quality parameters. Therefore, volunteer stream monitors are acting as extensions of the regulators, whether they collect routine grab samples or utilize data loggers, so they are able to observe and report pollution events to the proper regulatory agencies. (11-15-12 1, follow-up email)

Organizations that are monitoring the activities of the gas industry in ways that we refer to as "policing" do not necessarily seek to shut down or oppose gas drilling activities, but rather to facilitate official regulatory and enforcement processes. As the watershed association leader quoted above explained, "if there is a pollution event, DEP doesn't always have the staff to send immediately. They may get on site the next day, but by that time, the source could be washed downstream and diluted" (11-15-12 1). His organization's attitude toward the gas industry is not to "go out there and say, you are doing this wrong, you need to do it this way." Instead:

> ...The purpose of this effort is to get baseline data from streams that are near the well pads. ... If we can say this is our data, before, during and post drilling then hope to get some regulations in there, so that if gas companies are at fault for an incident they do whatever is necessary to fix it and prevent any future problems from occurring. We are not here to say we don't want the industry here, because we understand the necessity of natural gas extraction, we just want it to be conducted in the safest way possible in order to protect the health and safety of the environment as a whole. The gas industry is here and isn't leaving. We are monitoring waterways to protect aquatic and human life and to ensure that the industry follows existing environmental laws and regulations

> and will be held liable if an event should occur that can be traced back to their operations. (11-15-12 1)

Others are not so optimistic that their monitoring practices will elicit a response from regulators. One volunteer coordinator explained that she did not have much hope that the data loggers they were using would lead to a regulatory investigation, but saw other possible outcomes:

> Obviously if you download your data and you see a peak two weeks ago, I mean, it's long gone, DEP isn't going to do the investigation. So it's more, I think of the industry knowing that we have monitors out there and we are continuously watching what is going on. There is maybe some hope, we are not sure, down the line, how it will work, with litigation from the data from these as well, since it is maintained strictly by trained staff and individuals. You get volunteers and then the issue of litigation gets a little shaky. (8-2-12 1)

This volunteer coordinator indicated that having volunteers and data loggers in the field may increase the gas industry's efforts to avoid violations, because the likelihood of getting caught is increased. Echoing many others we interviewed, she went on to say: "I really think the fact that [the natural gas companies] know they are being watched pressures them to take a little bit more care than what they are doing [in other places]" (8-2-12 1). She also described the organization's monitoring program as important not only for potential use in future litigation, but also as a source of information that could be shared with the media or legislators. Her organization seeks not only to hold private companies accountable, but also to use their monitoring to keep government organizations like the DEP transparent: "we are going to hold them accountable as well as the gas industry."

For most of these organizations, litigation is not a primary strategy – only six organizations surveyed indicated that litigation was one aim of their water monitoring efforts – but in interviews, some project initiators mentioned the *possible* need for legal action if pollution is discovered and not addressed. However, it was sometimes noted that data collected by volunteers is unlikely to "hold up in court," which may be why few organizations listed litigation as an objective. Interestingly, the increasing use of continuous data loggers and the adoption of a science logic may make litigation a more viable option. Because data logger programs typically rely on standard technology and trained staff, rather than volunteers, the data may be accepted as credible in legal challenges. But replacing volunteers with data loggers might diminish the effectiveness of other forms of policing, like reporting problems the moment they are observed by volunteers.

An interview with two volunteer coordinators from an organization recently entering the field of shale gas monitoring revealed explicit and self-conscious experimentation with consciousness-raising, science, and policing logics. When asked about the benefits of having a volunteer stream monitoring program, the two speakers described multiple purposes for this work: raising public awareness, generating scientific baseline data, and reporting incidents to regulators with possible legal follow-up.

> Speaker 1: The public education and awareness is huge. Scientific – some background and familiarity with scientific fact is helpful in the activism world. How do people develop a relationship with the local agencies is really helpful. And then just having data on the ground to back up our activists is huge. ... So there are a lot of different factors involved. I think my primary interest, though, is baseline data, and we will use that.
>
> Speaker 2: Trying to legally force compliance with the law is an effective approach too. So it's this concept of having data available and being able to measure what is going on, and then being able to argue that you have been – you could bring your case to a court. I think is also a powerful message to send to the industry. ... [Y]ou may hear conflicting stories from us, because we are a small group and I think it's very much driven by personal experience, but – so I hope our story is consistent.
>
> Speaker 1: ... [O]ur model is just to identify an issue, report to the agency, enforce reporting and enforce action to the greatest extent possible using publicity, phone calls, etc. Awareness. (8-27-12 1)

This exchange illustrates that the logics of consciousness-raising, of science, and of environmental policing are not mutually exclusive. The plurality of logics employed by organizations, including the impulse to "force compliance with the law" are indicative of the complex organizational arrangements and competing missions developing across the field of water monitoring in the Marcellus Shale.

DISCUSSION

PEM is an increasingly common form of civic engagement, and many have attempted to theorize what it means in relation to other broad social changes. Is "citizen science" neoliberal or liberatory? Is it a form of participatory democracy or a "new tyranny" of public participation in power structures dominated by the agendas of scientific and political institutions (Cooke & Kothari, 2001)? We have argued that PEM resists generic characterization because it occurs in highly dynamic organizational fields, where "rolling turbulence" from changes in adjacent fields creates numerous opportunities for experimentation with new logics and practices. Field analysis offers a way to connect our knowledge of the broad shifts in environmental science and governance to the ethnographic knowledge that is generated through detailed case studies of participatory science. Within fields we find processes of collaboration and agenda-setting that not only determine the contours of particular PEM projects but also create new forms of interaction between citizens, science, and the state.

Environmental controversy is one source of opportunity for shifting these relationships and creating new forms of public participation. However, the creation of new participatory science practices may be constrained by established practices and objectives within civil society organizations. In this case, the participatory water monitoring field in the United States has historically been guided by a primary logic of consciousness-raising, demonstrated in practices that aimed to help volunteers to become more knowledgeable about their local watersheds, and thus more informed democratic citizens. With increasing demands for water quality data, both for government oversight and academic research, practices in the volunteer water monitoring field have, since the 1980s, also come to reflect a logic of science, not only educating volunteers, but also generating useful data for research purposes. The long-standing logics of consciousness-raising and science in the participatory water monitoring field – reflected in practices that prioritize public "awareness" and scientific data collection, rather than overt political confrontation – still shape the efforts to monitor the impacts of the gas industry.

The sudden threat of shale gas development, however, generated a new wave of turbulence, driving multiple changes in the interdependent fields of academic science, environmental regulation, and participatory water monitoring. Controversy over shale gas development led professional scientists to recognize the inadequacy of scientific understanding of the impacts of the industry, increasing and intensifying interactions between academic researchers and watershed groups. At the same time, watershed organizations identified the inability of government officials to adequately enforce environmental regulations, so they sought methods that would bring greater regulatory attention to the gas industry. New practices of data collection and sharing, reflecting the needs and expectations of academic science, were imported into the emerging shale impacts monitoring field, while environmental policing also became an orienting principle for at least some of the organizations in the field.

Some of these changes have had predictable outcomes. Shale impacts monitoring is increasingly scientized, reflecting the values and practices of the more dominant field of academic science. In particular, continuous data loggers, which ostensibly generate more reliable and complete scientific data, when compared to volunteer monitoring, are not particularly well suited to public education or environmental policing. They may reduce the participation of volunteers in favor of professional staff and they may delay the response to an acute pollution incident. For many, the adoption of a science logic means not engaging in direct advocacy against fracking. Initiators and volunteers frequently note that their influence derives from objectivity, which means not overtly taking political positions. On the other hand, there are likely to be unexpected outcomes of combining science and policing logics. For instance, it is possible, as at least one of our informants noted, that data loggers and sophisticated data management tools will enable organizations to shift to a strategy of environmental policing that focuses more on litigation rather than being merely the (often ignored) "eyes and ears" of government authorities.

Although each PEM field is likely to be different, we can draw some general lessons from our analysis of organizations monitoring the impacts of gas development. First, we observed that interactions between project initiators who draw on different institutional logics can bring about changes in practices. Interaction between academic scientists and watershed groups yielded new monitoring practices, such as greater attention to QA/QC and the compilation of measurements in shared databases. In other cases – such as air quality monitoring or drinking water monitoring – we might expect PEM organizations to interact with initiators from public health, social movement, educational, or even industrial fields, yielding new variations of PEM practices.

A second general lesson to be drawn from this analysis is that the impacts of macro-level changes, such as the scientization of politics, the rise of neoliberal policy ideas, or even large-scale industrial transformations as seen in the case of shale gas, are likely to be experienced in field-specific ways. The water monitoring field owes its long history of alignment with official watershed management programs to the wide-scale transformations in environmental governance that have been taking place since the 1980s. Organizations in other PEM fields, however, may be less affected by such policy changes, or may respond to them by taking an adversarial stance toward the state or scientific institutions. Each field is likely to be different; however, comparisons across PEM field are likely to reveal patterns that will provide more general insights about the forces that shape and constrain the practices and outcomes of participatory science.

NOTES

1. New York Department of Environmental Conservation (n.d.).
2. PA Department of Environmental Protection (n.d.).
3. To protect the confidentiality of speakers' identities, we use an interview identification system that includes the interview date followed by a unique number.
4. ALLARM (n.d.).
5. Dickinson College (n.d.).
6. Trout Unlimited (n.d.).
7. FracTracker (n.d.).
8. Community Science Institute (n.d.).
9. Pennsylvania League of Women Voters (n.d.).
10. One method of collecting samples of fish for analysis is to use an electrical current to stun them (see, e.g., USGS, n.d.).
11. Mountain Watershed Association (n.d.).
12. Mountain Watershed Association (n.d.).
13. Three Rivers Quest (n.d.).
14. CUAHSI (n.d.).
15. Colcom Foundation (n.d.).
16. Evergreen Conservancy (n.d.).
17. Loyalhanna Watershed Association (n.d.).
18. Khalequzzaman (2011).

ACKNOWLEDGMENTS

This material is based upon work supported by the National Science Foundation under Grant No. 1126235. The authors wish to acknowledge the contributions of Simona Perry to the early stages of this research. Elise Wilbourn assisted with coding and analyzing survey and interview data. They also wish to thank the organizers of and participants in the Political Sociology of Science Workshop, held in Madison, Wisconsin, in 2012, for offering insightful feedback on an early version of this paper. Many thanks are due to Scott Frickel and David J. Hess for their comments and guidance throughout the writing and revising process, and to Elizabeth Popp Berman, Kendra Smith-Howard, and Jennifer Dodge, who gave many helpful suggestions.

REFERENCES

ALLARM. (n.d.). About ALLARM. Retrieved from http://blogs.dickinson.edu/marcellus monitoring/about-allarm/. Accessed on March 3, 2014.

Beck, U. (1992). *Risk society: Towards a new modernity*. London: Sage.

Berman, E. P. (2012). *Creating the market university: How academic science became an economic engine*. Princeton, NJ: Princeton University Press.

Brown, P. (2007). *Toxic exposures: Contested illnesses and the environmental health movement*. New York, NY: Columbia University Press.

Cohn, J. P. (2008). Citizen science: Can volunteers do real research? *BioScience*, *58*(3), 192–197.

Colcom Foundation. (n.d.). *Regional interests*. Retrieved from http://www.colcomfdn.org/interests.html. Accessed on March 3, 2014.

Community Science Institute. (n.d.). *Our mission*. Retrieved from http://www.communityscience.org/?page_id=11. Accessed on March 3, 2014.

Cooke, B., & Kothari, U. (2001). *Participation: The new tyranny?* London: Zed Books.

Couch, S. R., & Kroll-Smith, S. (2000). Environmental movements and expert knowledge: Evidence for a new populism. In S. Kroll-Smith, P. Brown, & J. V. Gunter (Eds.), *Illness and the environment a reader contested medicine* (pp. 384–408). New York, NY: New York University Press.

CUAHSI. (n.d.). *Shale Network*. Retrieved from http://hiscentral.cuahsi.org/pub_network.aspx?n=228. Accessed on March 3, 2014.

DEP. (n.d.). *Outside agency data and quality assurance requirements*. Retrieved from http://files.dep.state.pa.us/Water/Drinking%20Water%20and%20Facility%20Regulation/WaterQualityPortalFiles/Methodology/OutsideAgencyData.pdf. Accessed on March 3, 2014.

Dickinson, J. L., Zuckerberg, B., & Bonter, D. N. (2010). Citizen science as an ecological research tool: Challenges and benefits. *Annual Review of Ecology, Evolution, and Systematics*, *41*(1), 149–172.

Dickinson College. (n.d.). *About ALLARM*. Retrieved from http://www.dickinson.edu/info/20173/alliance_for_aquatic_resource_monitoring_allarm/1528/about_allarm. Accessed on March 3, 2014.

Dunn, M. B., & Jones, C. (2010). Institutional logics and institutional pluralism: The contestation of care and science logics in medical education, 1967–2005. *Administrative Science Quarterly*, *55*, 114–149.

Entrekin, S., Evans-White, M., Johnson, B., and Hagenbuch, E. (2011). Rapid expansion of natural gas development poses a threat to surface waters. *Frontiers in Ecology and the Environment*, *9*(9), 503–511.

Evergreen Conservancy. (n.d.). *About EC*. Retrieved from http://www.evergreenconservancy.org/about/. Accessed on March 3, 2014.

Firehock, K., & West, J. (1995). A brief history of volunteer biological water monitoring using macroinvertebrates. *Journal of the North American Benthological Society*, *14*(1), 197–202.

Fischer, F. (2000). *Citizens, experts and the environment*. Durham, NC: Duke University Press.

Fligstein, N., & McAdam, D. (2011). Toward a general theory of strategic action fields. *Sociological Theory*, *29*(1), 1–26.

FracTracker. (n.d.). *Trout unlimited testing for water quality in PA's Marcellus region*. Retrieved from http://www.fractracker.org/2013/02/trout-unlimited-testing-for-water-quality/. Accessed on March 3, 2014.

Friedland, R., & Alford, R. (1991). Bringing society back in: Symbols, practices and institutional contradictions. In W. W. Powell & J. P. DiMaggio (Eds.), *The new institutionalism in organizational analysis* (pp. 232–263). Chicago, IL: University of Chicago Press.

Hess, D. J. (2007). *Alternative pathways in science and technology: Activism, innovation, and the environment in an era of globalization*. Cambridge, MA: The MIT Press.

Khalequzzaman, M. D. (2011). Water quality monitoring races natural gas development. *The Express*, March 10. Retrieved from http://www.lockhaven.com/page/content.detail/id/530063/Water-quality-monitoring-races-natural-gas-development.html. Accessed on March 3, 2014.

Kleinman, D. L. (2000). *Science, technology and democracy*. Albany, NY: SUNY Press.

Lave, R. (2012). Neoliberalism and the production of environmental knowledge. *Environment and Society: Advances in Research*, *3*(1), 19–38.

Lawrence, A. (2006). "No personal motive?" Volunteers, biodiversity, and the false dichotomies of participation. *Ethics Place and Environment*, *9*(3), 279–298.

Loyalhanna Watershed Association. (n.d.). *Water protection*. Retrieved from http://www.loyalhannawatershed.org/water.asp. Accessed on March 3, 2014.

Lynch, M. J., & Stretesky, P. B. (2013). The distribution of water-monitoring organizations across states: Implications for community environmental policing and social justice. *Policing: An International Journal of Police Strategies & Management*, *36*(1), 6–26. doi:10.1108/13639511311302452

McCormick, S. (2006). The Brazilian anti-dam movement: Knowledge contestation as communicative action. *Organization & Environment*, *19*, 321–346.

Moore, K. (2006). Powered by the people: Scientific authority in participatory science. In S. Frickel & K. Moore (Eds.), *The new political sociology of science* (pp. 299–323). Madison, WI: University of Wisconsin Press.

Moore, K., Kleinman, D. L., Hess, D., & Frickel, S. (2011). Science and neoliberal globalization: A political sociological approach. *Theory and Society*, *40*(5), 505–532.

Morello-Frosch, R., Brown, P., Altman, R. G., McCormick, S., & Mayer, B. (2006). Embodied health movements: Responses to a "scientized" world. In S. Frickel & K. Moore (Eds.), *The new political sociology of science* (pp. 244–271). Madison, WI: University of Wisconsin Press.

Mountain Watershed Association. (n.d.). *Shale gas & MCSP*. Retrieved from http://www.mtwatershed.com/shalegas/. Accessed on March 3, 2014.

Mountain Watershed Association. (n.d.). *Water quality monitoring*. Retrieved from http://www.mtwatershed.com/water-quality-monitoring/. Accessed on March 3, 2014.

National Science Foundation. (2011). *Can Marcellus Shale gas development and healthy waterways sustainably coexist?* Retrieved from http://www.nsf.gov/discoveries/disc_summ.jsp?cntn_id=122543. Accessed on January 2, 2013.

Nerbonne, J. F., & Nelson, K. C. (2004). Volunteer macroinvertebrate monitoring in the United States: Resource mobilization and comparative state structures. *Society & Natural Resources*, *17*(9), 817–839.

Nerbonne, J. F., & Nelson, K. C. (2008). Volunteer macroinvertebrate monitoring: Tensions among group goals, data quality, and outcomes. *Environmental Management*, *42*(3), 470–479.

New York Department of Environmental Conservation. (n.d.). Citizens Statewide Lake Assessment Program-Volunteers at Work. Retrieved from http://www.dec.ny.gov/chemical/79219.html. Accessed on March 3, 2014.

Ottinger, G. (2009). Buckets of resistance: Standards and the effectiveness of citizen science. *Science, Technology & Human Values*, *35*(2), 244–270.

Overdevest, C., Orr, C. H., & Stepenuck, K. (2004). Volunteer stream monitoring and local participation in natural resource issues. *Research in Human Ecology*, *11*(2), 177–185.

O'Rourke, D., & Macey, G. P. (2003). Community environmental policing: Assessing new strategies of public participation in environmental regulation. *Journal of Policy Analysis and Management*, *22*(3), 383–414.

Pennsylvania League of Women Voters. (n.d.). *About WREN*. Retrieved from http://wren.palwv.org/intro.html. Accessed on March 3, 2014.

Pfeffer, M., & Wagenet, L. P. (2007). Volunteer environmental monitoring, knowledge creation and citizen–scientist interaction. In *Sage handbook on environment and society* (pp. 235–249). London: Sage.

Rahm, B. G., & Riha, S. J. (2012). Toward strategic management of shale gas development: Regional, collective impacts on water resources. *Environmental Science & Policy, 17*, 12–23.

Sally, E., Evans-White, M., Johnson, B., & Hagenbuch, E. (2011). Rapid expansion of natural gas development poses a threat to surface waters. *Frontiers in Ecology and the Environment, 9*(9), 503–511.

Savan, B., Morgan, A. J., & Gore, C. (2003). Volunteer environmental monitoring and the role of the universities: The case of citizens' environment watch. *Environmental Management, 31*(5), 561–568.

Shirk, J. L., Ballard, H. L., Wilderman, C. C., Phillips, T., Wiggins, A., Rebecca, J., … Bonney, R. (2012). Public participation in scientific research: A framework for deliberate design. *Ecology and Society, 17*(2), 29.

Soeder, D. J., & Kappel, W. M. (2009). *Water resources and natural gas production from the Marcellus Shale*. US Geological Survey (pp. 1–6), May. Retrieved from http://pubs.usgs.gov/fs/2009/3032/. Accessed on June 7, 2013.

States, S., et al. (2011). Bromide in the Allegheny River and THMS in Pittsburgh drinking water: A link with Marcellus Shale drilling. In *American water works association, water quality technology conference*, Phoenix, AZ.

Stedman, R., Lee, B., Brasier, K., Weigle, J. L., & Higdon, F. (2009). Cleaning up water? Or building rural community? Community watershed organizations in Pennsylvania. *Rural Sociology, 74*(2), 178–200.

The Academy of Natural Sciences of Drexel University. (n.d.). A preliminary study on the impact of Marcellus Shale drilling on headwater streams. *Environmental Research*. Retrieved from http://www.ansp.org/research/environmental-research/projects/marcellus-shale-preliminary-study/. Accessed on January 2, 2013.

Three Rivers Quest. (n.d.). *What is 3 rivers quest?* Retrieved from http://3riversquest.org/about/3-rivers-quest/. Accessed on March 3, 2014.

Trout Unlimited. (n.d.). Trout unlimited provides stream surveillance training to Pa. Members in Marcellus Shale region. Retrieved from http://www.tu.org/press_releases/2010/trout-unlimited-provides-stream-surveillance-training-to-pa-members-in-marcellus. Accessed on March 3, 2014.

US Environmental Protection Agency. (1990). *Volunteer water monitoring: A guide for state managers*. Washington, DC.

USGS. (n.d.). *Shocking fish to collect biological samples*. Retrieved from http://ga.water.usgs.gov/edu/fishshock.html. Accessed on March 3, 2014.

Velinsky, D. (2010). *Testimony on the economic and environmental impacts of hydraulic drilling of Marcellus Shale on Philadelphia and the surrounding region*. Joint Committees on the Environment and Transportation & Public Utilities of the Council of the City of Philadelphia, September 23, 2010. Retrieved from http://catskillcitizens.org/learnmore/VELINSKYTEST.pdf. Accessed on March 3, 2013.

Ward, R. C. (1996). Water quality monitoring: Where's the beef? *Water Resources Bulletin of the American Water Resources Association, 32*(4), 673–680.

Wilson, D. (2002). Community based water monitoring and beyond, a case study: Pennsylvania. In *Proceedings of the Water Environment Federation* (pp. 1025–1036).